Open
Business
Models

Open
Business
Models

HOW TO THRIVE
IN THE NEW INNOVATION
LANDSCAPE

Henry Chesbrough

HARVARD BUSINESS SCHOOL PRESS

Boston, Massachusetts

Library of Congress Cataloging-in-Publication Data
Chesbrough, Henry William.
    Open business models: how to thrive in the new innovation landscape /
Henry Chesbrough.
        p.   cm.
    ISBN13: 978-1-4221-0427-9
    ISBN10: 1-4221-0427-3
    1. Technological innovations—Management.   2. Research, Industrial—
Management.   I. Title.
    HD45.C468 2006
    658.4'063—dc22
                                                              2006030362

The paper used in this publication meets the minimum requirements of the
American National Standard for Information Sciences—Permanence of Paper
for Printed Library Materials, ANSI Z39.48-1992.

For Emily and Sarah

Contents

Acknowledgments

This book results from listening to and learning from many people in both industry and academia who responded to my previous book, *Open Innovation*. That book argued that companies should use external ideas more in their own businesses and allow their unutilized ideas to be used by others. I have learned that this works on a personal level as well. Many of the ideas in this latest book originated from discussions with different people, while others were refined and improved. I hope that this book does justice to these insights by combining them with my own and producing something of greater value for all of us.

Many of my colleagues at the University of California, Berkeley, have made substantial contributions to the work contained in this book. Within Berkeley's Haas School of Business, I have benefited from the thoughts of David Teece, Robert Cole, Jerry Engel, Bronwyn Hall, Drew Isaacs, Ray Miles, David Mowery, and Jihong Sanderson. Many Berkeley students have provided excellent research assistance, including Elsie Chang, Alberto Diminin, Helen Liang, Xiaohong Quan, Cengiz Ulusarac, and Simon Wakeman. Outside of Haas, in the surrounding Berkeley academic community, I have also held very helpful meetings with Robert Glushko, Robert Merges, Rhonda Righter, Annalee Saxenian, and Pam Samuelson.

Not all of the smart people in the world work at Berkeley, however. Two academic scholars who have tested these ideas and contributed to them include Joel West and Wim Vanhaverbeke, my coauthors on another, more academic text: *Open Innovation: Researching a New Paradigm* (Oxford University Press, 2006). Other academics who have helped me think through many of the issues in this book include

Melissa Appleyard, Ashish Arora, Jens Froeslov Christensen, Oliver Gassmann, Michael Jacobides, Peter Koen, Keld Laursen, Kwanghui Lim, Ikujiro Nonaka, Gina O'Connor, Andrea Prencipe, Richard Rosenbloom, Ammon Salter, Stefan Thomke, Chris Tucci, Max von Zedtwitz, and Rosemarie Ziedonis.

Another critical source of information for this book has come from managers of companies struggling with the innovation process in their own firms. Many of these people are identified and quoted within the text of this book, and I won't lengthen this acknowledgment by repeating all of those names. Special help and advice that went beyond the call of duty, however, does deserve special mention: Alpheus Bingham of Eli Lilly; Julie VanDerZanden of K2 Sports; Gil Cloyd, Martha Depenbrock, Larry Huston, and Jeff Weedman of Procter & Gamble; Ed Kahn and Michael Kayat of EKMS; John Amster, Peter Detkin, Nathan Myhrvold, and Laurie Yoler of Intellectual Ventures (Laurie Yoler left Intellectual Ventures while this manuscript was in development); Jim Huston and David Tennenhouse, both formerly of Intel; Michael Friedland and John King of Knobbe Martens; John Wolpert of the InnovationXchange; Joel Cawley, Mark Dean, Paul Horn, Jean Paul Jacob, and Jim Spohrer of IBM; Suzanne Harrison and Pat Sullivan of ICMG; and Eric Hahn, formerly of Collabra and Netscape. Despite all of their help and feedback, there are undoubtedly still many errors in this book. However, they are new and better errors than I would have made, had I not talked to these people.

My friends Rich Mironov, Ken Novak, Arati Prabhakar, and Pat Windham all patiently endured my early attempts to explain the concepts in this book. I am also indebted to my editor, Jeff Kehoe, at Harvard Business School Press for his support and guidance through the manuscript's development and editing process.

The students in my classroom have been a vital part of my own process of reflection on industrial innovation. While they do not yet have the years of experience that my managerial sources possess, they bring a fresh perspective that challenges the conventional wisdom that often accompanies deep experience. Their questions, arguments, and conclusions have helped me test and revise my own thinking about innovation.

My wife, Katherine, again read through the entire manuscript and painstakingly exposed the gaps, errors, and incomplete thinking of earlier drafts. The book is much, much better for her patient reading. I am also indebted to my parents, Richard and Joyce Chesbrough, for their support in writing this book.

This book is about building a more open and better future for our companies and ourselves. My daughters, Emily and Sarah, are my two proudest "innovations." Each is very open with me (whether I want to hear it or not), and each promises to make a wonderful contribution to the world of tomorrow. I dedicate this book to them as a proud parent.

—Henry Chesbrough
Berkeley, California
chesbrou@haas.berkeley.edu

Preface

Everyone knows that innovation is a core business necessity. Companies that don't innovate die. This is not news.

In the current environment, however, to innovate effectively, you increasingly must innovate openly. And to innovate openly, you must do more than search externally for new ideas or license out more of your own ideas. You also must innovate your business model, the way that you create value, and capture a portion of that value for yourself. This *is* news. This job cannot be delegated to the head of R&D or the chief technology officer. Business model innovation is the job of every senior manager, especially those in general management, marketing, business development, legal (particularly intellectual property), finance, and new product and process development.

Innovation is often expensive, but consider how expensive it would be to stop innovating. The key is to make innovation more effective in cost, time, and risk management by extending it to business models. This book describes how organizations can thrive in an era of Open Innovation. Open Innovation means that companies should make much greater use of external ideas and technologies in their own business, while letting their unused ideas be used by other companies. This requires each company to open up its business model to let more external ideas and technology flow in from the outside and let more internal knowledge flow to the outside. With a more open business model, Open Innovation offers the prospect of lower costs for innovation, faster times to market, and the chance to share risks with others.

These benefits, however, also entail costs. The more companies learn about this idea, the more they realize how much they have to change their own innovation activities to take full advantage of it. As with any significant change, there are real barriers to overcome in order to make Open Innovation work.

This book describes the current innovation landscape and focuses on the changes required to make Open Innovation happen. These go well beyond a company's innovation process to its business model. To thrive in this era, companies have to adapt business models to make them more open to external ideas and paths to market. Those that effectively bring ideas from the "outside in" will tap into tremendous potential for identifying and creating new value; likewise, companies that move ideas from the "inside out," enabling others to use unused ideas, will realize a new way to capture more value and sustain themselves in these times of increasingly global markets and competition.

This book does not assume familiarity with my earlier book, *Open Innovation* (although I recommend it to you!). I have written *Open Business Models* both to inform new readers who are not familiar with Open Innovation about its power and value, and to instruct readers who know about Open Innovation on how to take the next steps to make it pay off.

The book starts in chapter 1 by examining why Open Innovation makes so much economic sense in a world of widely distributed knowledge, emerging intermediate markets for that knowledge, and ever-shortening product lives. Markets for knowledge have existed for a long time; however, they have been highly inefficient. What is changing is that information technologies have reduced some of these inefficiencies and have prompted newly entering companies to perform roles that make a market for ideas more feasible. We examine how markets for ideas are driving innovation in the chemicals industry, the pharmaceutical industry, and the entertainment industry.

If chapter 1 describes the opportunity presented by more open business models, chapter 2 explores their dark side. It is not easy to embrace an open business model. There are significant barriers and costs that an open business model is likely to encounter inside most companies. Behaviors such as the "not invented here" syndrome inhibit a company's ability to search for external sources of ideas that could advance the business model. And the "not sold here" virus keeps potentially valuable internal ideas bottled up inside the firm.

Building and growing a strong innovation system requires overcoming these barriers. Open source software provides one example where openness is creating significant value, a portion of which is captured by the firm, via a business model built around open source.

In chapter 3, we explore one of these barriers, intellectual property (IP), both historically and in the context of innovation. Here we also discuss the concept of secondary markets for innovation. The two concepts are related, because IP protection supports secondary markets to trade that IP. While this trend is in its infancy, some interesting qualitative evidence shows that these markets are emerging. Some quantitative data from patent reassignments further supports the growing extent of these markets.

Chapter 4 develops a more conceptual approach to IP management. IP's ability to protect technology is uneven. And technology itself must be managed through a technology life cycle. There are four phases in the cycle: emerging, growth, maturity, and decline. IP should be managed differently in the phases of that life cycle and linked to the company's business model.

The second half of the book shifts from the more conceptual material of chapters 1 through 4, to more practical examples and frameworks for how to implement open business models.

In chapter 5, the concept of an open business model is developed, along with a six-stage framework to advance your business model. The business model framework can serve as a benchmark for companies to assess their current business model and determine the appropriate next steps to advance it to a higher stage of openness. A variety of diagnostic questions help develop an initial assessment and also point the way toward more advanced innovation processes.

Chapter 6 introduces us to a new set of players in the innovation market, a group of organizations I term *innovation intermediaries*. These players owe their existence to the emerging market for IP, and their business models are creating far greater access to a global community of innovation providers. These companies illustrate some emerging practices for bringing external ideas inside and/or taking internal ideas outside.

Chapter 7 discusses companies that have built their own business models around IP, often to enable others to create value with that IP, other times to capture value from those that may infringe on that IP. Some of these IP-based models are quite benign; others are potentially more predatory. It seems clear, though, that IP-based models are growing in importance and that they will strongly influence the innovation activities of companies in many ways.

Chapter 8 closes the book with an extended discussion of how three very different companies—Air Products, IBM, and Procter & Gamble—have altered their business models to become much more open. These companies help point the way for others that wish to follow. Making this transformation will require new processes, new metrics of performance, a clear-eyed view of your own business model, and changing some long-held views about innovation and IP.

Innovation is an extremely dynamic activity to study, by its very nature. Ideas and practices can become obsolete very quickly. Once you have read the book, please visit the Web site http://openinnovation @haas.berkeley.edu (the U.S. site) and/or www.openinnovation.eu (the European site). There you can observe new applications of these ideas and register to receive additional updates. As you might expect from people interested and involved in Open Innovation processes, these sites are not restricted to just one person's ideas, but are meant to serve as a resource for a community of people who believe that they can innovate more globally and more effectively by sharing ideas and results with each other.

1

Why Business Models Need to Open Up

Former baseball catcher Yogi Berra liked to observe that "the future isn't what it used to be." This same general pessimism has been voiced about innovation in the United States and Europe, even as the innovation opportunities globally in places such as China and India have greatly expanded.[1]

And it is not hard to see why. The great research and development laboratories of the twentieth century have been downsized, broken up, or redirected to new purposes in the West, while new labs are springing up overseas in large countries such as India and China, as well as small countries such as Finland and Israel. Companies are shortening their time horizons for research and development expenses and are shifting money from "R" to "D." Many great universities are putting up patent fences around new research discoveries that are funded by taxpayer dollars. Start-up companies once were the great hope that would fill the void in corporate innovation, but the bust in the Internet sector and the venture capital market has dampened these expectations. Fears of outsourcing high-wage jobs to low-cost centers around the world dominate many discussions of economic and trade policy. In many varied parts of the western innovation system, the future of innovation has clearly changed.

Even the rapid emergence of India and China into the international economic system does not assuage these concerns. Transferring innovative activities to more efficient regions saves money in the short run but by itself may not do much to increase the number of new ideas that are needed to advance technology and improve our standard of living in the long run. Where will the critical breakthrough ideas of the twenty-first century come from? These changes raise concerns about the ability of advanced economies to continue to innovate at the pace of the last century.

But hidden among these worrisome trends are other developments that are perhaps more hopeful for the future of innovation. One such development is the growing division of innovation labor. By a "division of innovation labor," I mean a system where one party develops a novel idea but does not carry this idea to market itself. Instead, that party partners with or sells the idea to another party, and this latter party carries the idea to market. This new division of labor is driving a new organizational model of innovation, one that may offer more hopeful prospects for innovation in the future.

To tap into this new division of labor, companies will have to open up their business models. If they are able to do so, many more ideas will become available to them for consideration, and many more pathways for unused internal ideas will emerge to unlock latent economic potential as those ideas go to market. Companies that effectively build or change to open business models to exploit these opportunities are likely to prosper.

Let's be clear about what is meant by an *open business model*. A business model performs two important functions: it creates value, and it captures a portion of that value. It creates value by defining a series of activities from raw materials through to the final consumer that will yield a new product or service with value being added throughout the various activities. The business model captures value by establishing a unique resource, asset, or position within that series of activities, where the firm enjoys a competitive advantage.

An open business model uses this new division of innovation labor—both in the creation of value and in the capture of a portion of that value. Open models create value by leveraging many more ideas, due to their inclusion of a variety of external concepts. Open models can also enable greater value capture, by using a key asset, resource, or position not only in the company's own business but also in other companies' businesses.

To understand the attractive potential of this new approach to innovation, and the role of an open business model, consider the following names and what they may have in common:

Qualcomm, a maker of cellular phone technology

Genzyme, a biotechnology company

Procter & Gamble, a consumer products company

Chicago, the musical stage show and movie

This assortment of companies and products might strike you as random, but all share a common characteristic: each started with an idea that traveled from invention to market through at least two different companies, which divided the work of innovation between them. Qualcomm, for example, used to make its own cell phones and base stations but stopped doing that years ago.[2] Others make those products now. Today, Qualcomm makes chips and sells licenses to its technologies, period. Every phone using its technology is sold by a customer of Qualcomm, not the company itself.

Genzyme has achieved its success by licensing technology in from outside the company and then developing that technology further within the company. It has developed these external ideas into an array of novel therapies that deliver important cures for previously untreatable, rare diseases. It has also built a record of impressive sales and profits in an industry where profits have been hard to obtain.[3]

Procter & Gamble (P&G) has rejuvenated its growth through a very successful program called Connect and Develop, which licenses in or acquires products from other companies (such as the SpinBrush, Olay Regenerist, and the Swiffer) and takes those products to market as P&G brands. The company now actively seeks external ideas and technologies through an extensive network of scouts.

Chicago, the often-revived musical stage production, emerged out of a creative extension of a play written long ago, which was out of print.[4] Others saw the latent value within this play and revived it multiple times to become a prize-winning show. And each time the show was revived, it was done by a different owner! The most recent revival, in 1997, turned into an Academy Award–winning movie in 2002 and even created a best-selling musical soundtrack in the process.

There is something else that these diverse items all share: all were assisted by the astute management of an open business model. These items changed hands at least once in their journey to the market: they were bought, sold, licensed, or otherwise transferred from one party to another. Without a business model that sought to harness these external parties (and without the effective management of intellectual property to work with them), the resulting innovations may never have seen the light of day.

If these ideas were so valuable, an obvious question arises: why didn't the original owners of them figure out the best way to use them and take them to market? The answer goes to the very heart of what is important about intermediate markets for innovation. Different companies possess different assets, resources, and positions, and each has a different history.[5] These differences cause companies to look at opportunities differently. Companies will quickly recognize ideas that fit the pattern that has proved successful for them before. But they will struggle with ideas that require a very different configuration of assets, resources, and positions to be successful. With intermediate markets, ideas can flow out of places where they do not fit and find homes in companies where they fit better. So let's examine these intermediate markets.

INTERMEDIATE MARKETS FOR IDEAS AND TECHNOLOGY

Intermediate markets are markets in which an upstream supplier licenses its know-how and intellectual property to downstream developers and producers.[6] In intermediate market situations, different ingredients for business success (the idea itself, the critical development, manufacturing and distribution assets, the intellectual property [IP]) may all lie in different hands.[7] As intermediate markets grow in importance (in industries ranging from software, semiconductors, telecommunications, industrial chemicals, and petroleum engineering, to name a few), an increasing number of situations have emerged wherein the owner of the idea or the technology, the owner of the key business assets, and the owner of the requisite intellectual property all differ.

One advanced example of the variety of players and assets that have to be combined for business success comes from Hollywood. In the entertainment industry, scriptwriters shop their screenplays, the producers raise and supply the money, actors and directors are hired, a set location is identified or built, and special effects subcontractors are retained, along with the camera operators, the crew, and the catering company. Every major movie is a project, created from a unique configuration of these players. And everyone has an agent.

These different players—each with unique assets, resources, and positions—influence these markets for ideas and technology. This influence, in turn, affects the ability of innovators to capture value from their innovation investments. In the old Hollywood studio days, all of the major activities to develop, make, and distribute a movie were performed by the studio. The business model created value by offering movies that were popular and captured value by keeping its top resources (actors, directors, movie houses) under exclusive contracts.

Today, with the rise of intermediate markets, the question of crafting effective business models is more nuanced and intriguing than in the days of the Hollywood studio. This is because the presence of many external players yields many potential alternative models to consider when developing a movie. Should this script be shopped around, to see who is most interested? Can we attract cofinancing? Can we get an A-list director to make the film? Have any major stars expressed interest in the project? Should we roll out all at once, or start more slowly, or just release straight to video? Today's movie business models must answer all of these questions, with no two movie projects answering them in exactly the same way.

INEFFICIENCIES IN MARKETS FOR TECHNOLOGY

Markets for upstream technologies have existed in many industries for a very long time. Historically, however, in most industries they have been highly inefficient. The information that firms need to transact with one another was lacking. Information on what was available at what price was costly to obtain, and was not widely shared as a result. Even now, much of the exchange of technology and its associated IP happens through a cottage industry of brokers and patent attorneys. While transactions do occur, the price and other terms of the transactions are difficult to discern. Licensing of technologies from one company to another happens more frequently than is usually recognized, but licensing is hard to document because it is seldom reported on most corporate financial disclosures. This makes it hard to determine the overall amount of activity and a fair price for the technology.

When intermediate markets are highly inefficient, innovations and IP are exchanged in other ways. Companies buy and sell other companies to get access to the underlying ideas and technologies. Spin-offs and breakups represent different mechanisms to allow firms the ability to gain access to specific intellectual property that is then placed in the new venture. These may not be the ideal ways to access innovation and IP assets, because many other assets come along with these transactions, but at least they provide some means to access external technology. We will consider these intermediate markets in greater detail in chapter 3.

And of course, a good deal of potentially valuable trade in innovation and its associated IP does *not* occur in highly inefficient markets. The costs are so high, and the potential value so hard to perceive, that innovation and IP often sit on the shelf unused. One way to measure this is to quantify the utilization rate of patents owned by a company. This is a measure of the percentage of patents that are used in the firm's business, divided by the number of patents owned by that firm. In an informal survey I have done, I find that companies use less than half of the patented technologies that they own in at least one of their businesses. The range I have heard is between 5 percent and 25 percent, meaning that in this unscientific sample, somewhere between 75 percent and 95 percent of patented technologies simply lie dormant.

This is an extraordinarily wasteful state of affairs from a shareholder viewpoint. Shareholders are funding a company's R&D in order to produce valuable technologies that can contribute to the company's success in the market. Patented technologies that are not used in the company's business, and are not used by anyone else, are a waste of shareholders' money.[8]

THE SOCIAL COMPACT UNDERLYING
IP PROTECTION

The waste is even greater from a societal perspective. The patent system is a social contract between society and its inventors. Society wants to elicit brilliant inventions from its citizens. It dangles a temporary monopoly in front of the inventor to induce him or her to perform the arduous work and take the financial risks necessary to create new technologies and carry them to market. But the inventor can only receive a patent on the invention if the inventor discloses the invention in sufficient detail that others "practiced in the art" could also make the invention. So society offers a temporary monopoly, with the prospect of riches that this may entail, in return for disclosure of the invention. This disclosure also enables other inventors to build on the earlier invention upon expiration of the patent so that the technology can be improved over time.

If, however, the inventor never practices the technology he or she patented, or never licenses others to practice the technology, then the product covered by the temporary monopoly is never taken to market. Not only does society lose the use of the new invention; society has empowered the inventor to prevent anyone else from using it until the patent expires.[9] While there is no legal requirement that the inventor receiving a patent make some practical use of it, the logic behind the social contract is broken when inventors do not use their inventions or license others to use these inventions. When corporations build enormous patent portfolios amounting to thousands or tens of thousands of patents and use less than 25 percent of them in any of their businesses, the social loss is substantial.

Other forms of intellectual property protection, such as copyright, trade secrecy, and trademarks, offer slightly different versions of this social contract. All seek to promote the advancement of commerce and technology by creating incentives for innovation investment, and each provides a certain level of social protection for owners of the IP in order to do so. And this compact that provides private protection in return for societal advancement goes back to the founding of the U.S. patent system, as we shall see later.

CLOSED IP MANAGEMENT

Until recently, IP was managed by most companies as a backwater activity. Most companies delegated the management of their IP to legal specialists, such as an in-house attorney or an outside patent counsel. These specialists were unlikely to be able to (or even allowed to) connect IP to the company's overall business model and innovation process. Their primary job was to keep the company out of legal trouble. These specialists only had to worry about what could go wrong. Their role was not designed to create competitive advantage for the company. That role was given to the R&D organization and the business units that used its outputs.

And these functional areas of the company were generally closed, from an IP point of view. The IP was created internally, used internally, and brandished only on occasion externally to ward off intruders or settle an outstanding litigation claim. External intellectual property was regarded as suspect, unreliable, and something to be avoided.

Consider the pharmaceutical industry up through the 1980s. All of the leading companies pursued extensive internal programs of discovery and development to create new drugs. Whenever a promising compound was identified, the pharmaceutical company would file for one or more patents on that compound. A select number of these compounds would be chosen for clinical development, and a very, very select few actually made it through to the market. But for every compound that turned into a new drug, there were scores of other compounds that were patented, pursued, and then eventually abandoned. These abandoned compounds, and the IP that protected them, were simply placed on the shelf. This practice was reckoned to be a regrettable but unavoidable cost of doing business.

This phenomenon is hardly unique to pharmaceuticals. In Germany, an executive at Siemens informed me that more than 90 percent of all German patents issued are never commercialized in the market. (This was also apparently true of Siemens's use of its own patents.) Procter & Gamble in 2002 estimated that it used only about 10 percent of the patents it held.

This behavior is what I term "closed IP management": there is only one way to access the IP (i.e., from within your own firm) and only one way to deploy it (through your own products selling to the market). In this model, the majority of IP never gets used. In chapter 2, we will explore some reasons for this low rate of utilization of ideas and their associated IP.

Companies differ in their response to the low utilization rate of their ideas and IP. When told about the fact that, before 2002, 90 percent of P&G's patents were not being used anywhere in any of its businesses, companies offer one of two responses. Both are quite revealing. The first response is a knowing chuckle, with an acknowledgment that the situation is very similar in their own company. The second response is a look of shock, as the person realizes that *no one in the company has ever asked the question*: nobody actually knows what proportion of IP assets are actually used in the company's business.

This low utilization rate is not surprising when one considers how the top IP managers in the company are compensated. Rarely do companies link the compensation of IP managers to their ability to create financial returns on those assets. Most IP managers are rewarded for keeping the company out of trouble, not for using those IP assets to contribute to the financial performance of the organization.

As you might expect, to ask this question is to begin the process of changing the answer to it. P&G has taken significant steps to increase its patent utilization. Siemens has also embarked on a program to rapidly increase the utilization of its patents. A few other companies have started to address the lack of management of this class of corporate assets. The vast majority, I suspect, have yet to assess their position, which means that any greater degree of effective management is even further off in the future. When they do make that assessment, they will find that they will have to open up their business models to make more effective use of their IP assets.

MANAGING IP FOR VALUE CREATION

Many executives, when they think about managing IP, think of it solely as a means to extract value from a technology or a set of technologies. While IP can be used this way, that is only part of its importance. Firms developing new technologies and new products pursue IP protection primarily for defensive reasons, to ensure their ability to practice their technology in their business without fear of interruption. The presence of patents becomes an insurance policy against unwelcome litigation and acts as a powerful bargaining chip in situations where litigation arises.[10]

Even that approach, however, is insufficient in a world of Open Innovation. IP cannot be used to extract value unless and until there has been some value created by the technology or technologies. As we shall see, IP can be managed to help create value, not simply capture it, particularly when its management is linked to the company's business model and innovation process. For example, companies might choose to publish, or give away a portion of their IP, to create standards or to establish an intellectual commons—a safe harbor for development where knowledge is held in common—that will foster useful advances that in turn can enhance their own business. This is a dimension that is too often ignored in the texts that discuss how to manage IP or how to design social policies to optimize IP protection.[11]

WHY OPEN UP NOW? THE CHANGING ECONOMICS OF INNOVATION

Why open up your business model, tap into the intermediate markets for ideas and innovations, and make the associated changes to manage your IP? One part of the answer was given in chapter 3 of *Open Innovation*.[12] That chapter showed how useful knowledge and technology was becoming increasingly widespread, distributed across companies both large and small, in many parts of the world. There is simply too much good stuff out there for even the best companies to ignore.

Here we will focus on some specific forces in the economics of innovation that are forcing companies to open up their innovation process. These forces include the rising costs of technology development, combined with the shortening of shipping lives of products. Together, they make R&D investment under the closed model of innovation increasingly difficult to sustain.

Rising Costs of Technology Development

One example of the rising costs of technology development is the cost of building a new semiconductor fabrication facility (known as a "fab"). In 2006, Intel announced it would build two new fabs, one in Arizona and one in Israel. Each facility is estimated to cost more than $3 billion. Twenty years ago, a new fab would have cost about 1 percent of the amount of these new facilities. Another example is pharmaceutical drug development, which has risen to well over $800 million for a successful product, up more than tenfold from just a decade earlier. Even in consumer products, P&G estimates that its Always brand of feminine hygiene pads cost $10 million to develop a decade ago. According to P&G's Jeff Weedman, a similar development today would cost P&G between $20 million and $50 million.

Shorter Product Life Cycles

By itself, rising costs of technology development might mean that the big will get bigger and that everyone else will fall behind. But a second force makes these economics challenging even for the largest firms: the shortening product life cycles of new products today. In the hard disk drive industry where I used to work, our products in the early 1980s would typically ship for four to six years, once we achieved a "design win" to have a manufacturer use our drives. By the late 1980s, the expected shipping life fell to two to three years. By the 1990s, the expected shipping life was six to nine months. After that, a new and even better product was available.

In pharmaceuticals, the expected shipping lives of new drugs while they enjoy patent protection have also shortened. FDA approval now takes eight to ten years for typical drugs. And an army of over-the-counter pharmaceutical companies will copy any successful drug as soon as it comes off patent. In the largest market segments, successful drugs now also must share the market with rival patented drugs, even while the patents are still in effect.

Anyone who has purchased a cell phone in the past year can vouch for how quickly the product life cycles are moving in that market. New phones are coming into the market every six months, and a phone purchased just two years ago can feel surprisingly obsolete. Now phones include cameras, video, and connectivity to the Internet, to name a few of the most salient features that have been added.

The combination of rising development costs and shortening market windows compresses the economics of investing in innovation, reducing the company's ability to earn a satisfactory return on its innovation investment. Figure 1-1 illustrates this change. In this figure, the "closed model—before" shows expected revenues far in excess of the development costs. As development costs rise and as the market lives of the offerings becomes shorter, however, the net result is that the "closed model—after" finds it harder to justify the innovation investment.

FIGURE 1-1

The economic pressures on innovation

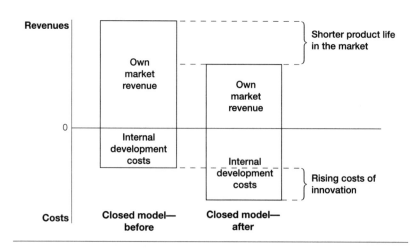

A more dynamic way to think of this is to compare the growth rate of R&D expenses to the growth rate of sales within an industry. Two curves that are growing at very similar rates suggests that the current business model is sustainable. If, however, the R&D expense curve is growing at a faster rate than the sales curve, that growth in R&D expense is clearly unsustainable. Without a change in the business model, a dramatic reduction in R&D expenses is inevitable.

Figure 1-2 shows an analysis of these curves for the pharmaceutical industry. (I am indebted to Alpheus Bingham of Eli Lilly for this data.) Note that the sales data corresponds to the left vertical axis, while the R&D data corresponds to the right vertical axis in the figure.

As the curves show, the sales of the pharmaceutical industry grew very nicely in the past twenty-five years, with an annual compound growth rate of about 11 percent. However, the R&D cost curves have grown even faster during this period, rising at about a 15 percent compounded rate of annual growth. This suggests that the economics of innovation within the pharmaceutical industry have become increasingly unsustainable. The blockbuster business model of screening thousands of compounds for a single drug that yields sales of $1 billion or more is getting hammered both by rising development costs, and by increased competition as other companies'

FIGURE 1-2

Pharmaceutical sales growth versus R&D growth

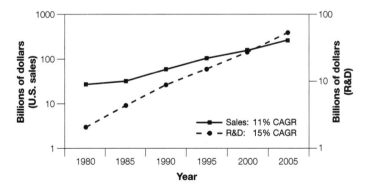

Source: Alpheus Bingham of Eli Lilly, PhRMA.

own blockbuster drugs come out in the very same market categories—effectively shortening the economic life of the original blockbuster. Older drugs also face strong competition—in the form of
generic drugs.

EARLIER INITIATIVES TOWARD
MORE OPEN BUSINESS MODELS

Others have stated the case for managing the company's assets in a
more open way. A particular thrust occurred in the late 1990s as
companies began to open up the licensing of their technologies to
other firms. A group of managers formed the Licensing Executives
Society to exchange ideas and best practices in licensing out patents
and other IP. Commentators like Petrash, Sullivan, and especially
Rivette and Klein called attention to the profit opportunity latent in
intellectual property.[13] Rivette and Klein's effort, titled *Rembrandts
in the Attic*, promised great riches to those who, as the title implied,
would dust off their IP, bring it down from the corporate attic, and
offer it for sale to others.

These efforts provided useful facts to managers and executives
charged with leading this activity, allowing companies to benchmark
their IP management and licensing activities against those of other
firms. This benchmarking was helpful, in that it established what
other companies were doing. A wave of consulting practices emerged
to help companies value their IP and prepare it for sale. Others developed software to increase the ability to detect infringement, and
still others offered expertise to aggressively assert IP owners' claims
against those that infringed on their patents. Web sites were created
to establish marketplaces where IP could be traded. It seemed that
intermediate markets for IP were well on the way.

Yet all this work ultimately ended in disappointment. Most of
the boutique consultancies that promulgated IP benchmarking practices have closed their doors. The internal practice areas launched
inside of the big accounting firms similarly have folded. The Web
sites that were going to create online auction sites for patent licens-

ing have transformed into something else or are now defunct.[14] The valuation methodologies of the consultants that attempted to determine the value of IP also have generally been discredited.

What was lacking? There was no connection between the management of IP and the innovation process that created this IP. Further, there was no linkage between the IP management and the business model of the company. For example, there was no consideration of when companies should outlicense technology to others and when they should refrain from doing so. Benchmarking could not answer these questions, because it identified company practices but could not probe the logic underlying those practices.

An obvious logical gap was the lack of a two-sided market for IP. It takes buyers as well as sellers to create a market. Sure, companies might wish to *sell* their IP, but why would anyone ever want to *buy* it from them? IP assets were not "Rembrandts," at least not to most companies. If the selling company did not want to use its IP, why would a buyer want to use that IP? What value is there for these "leftovers"?

There needed to be a rationale for companies to want to buy someone else's IP, if there was to be a market for these assets. And that rationale had to value the IP from the viewpoint of a willing buyer, not just that of an eager seller. Open Innovation provides a clear rationale for participating in intermediate markets. In a world of widely distributed useful knowledge, one can only sustain innovation by actively licensing in external ideas and technologies alongside developing and deploying one's own ideas. One can further justify the active outlicensing of internal ideas that are not being used to other firms.

The Open Innovation perspective is why IP management will be received differently this time. The earlier approach to IP management did not appreciate the need to link IP management to one's innovation processes and one's business model. A new kind of open business model and an open approach to managing IP are required to sustain and advance the company's business in a world of abundant knowledge. This model links the growth in markets for ideas with innovation processes to develop and advance business models.

OPEN BUSINESS MODELS: ADDRESSING
INNOVATION'S COSTS AND REVENUES

As we saw in figures 1-1 and 1-2, the economics of innovation are being negatively impacted by rising innovation costs and shorter revenue streams. Open business models address both issues. The Open Innovation business model attacks the cost side of the problem by leveraging external R&D resources to save time and money in the innovation process. One powerful testament to both cost savings and faster time to market can be found in an article by Larry Huston and Nabil Sakkab of Procter & Gamble.[15]

There, two P&G executives document the significant cost savings and time savings that P&G has realized from leveraging external technologies. The Pringles Print initiative is one excellent example of these savings. P&G now offers Pringles chips with pictures and words printed on each chip. The company found a bakery in Bologna, Italy, that had an ink-jet method for printing messages on cakes and cookies, which P&G adapted to Pringles. P&G got the product developed at a fraction of the cost and got it to market in half the time that the project would have taken internally.

The Open Innovation model attacks the revenue side by broadening the number of markets addressable by the innovation.[16] (That same P&G article by Huston and Sakkab illustrates this as well.) P&G is creating new brands from licensing in technologies from other companies around the world, resulting in products like the Crest SpinBrush, Olay Regenerist, and Swiffer dusters. P&G is also getting money from licensing its technologies to other companies.

The interplay of leveraged cost and time savings, combined with new revenue opportunities, is shown in figure 1-3. In this figure, the firm no longer restricts itself to the markets it serves directly. Now it participates in other segments through licensing revenues, joint ventures, spin-offs, or other means. These different streams of revenue create more overall revenue from the innovation. Meanwhile, the development costs of innovation are reduced by greater use of external technology in the firm's own R&D process. This saves time, as well as money. The result of the model is that innovation becomes economically attractive again, even in a world of shorter product life cycles.

FIGURE 1-3

The new business model of Open Innovation

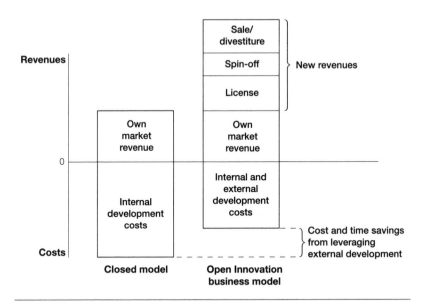

POSSIBLE OPPORTUNITIES AND THREATS FROM INTERMEDIATE MARKETS

There is growing evidence that intermediate markets are becoming an integral part of innovation and growth. While the discussion so far has been very positive about this development, there are real risks that arise as well. These intermediate markets can help you enter into new businesses, but if you don't manage them well, these markets may be used by others to *block* you from entering into new businesses, or to tax you in businesses in which you already participate. Intermediate markets can provide more revenue sources for important discoveries that you make. However, these markets may be used by others to impair your ability to use the discoveries you make. These markets will usher in new kinds of open business models. These markets may destroy some long-established business models. Whether these intermediate markets will help you, or hinder you, will depend on your ability to manage them and the business model you create for them.

If you manage intermediate markets skillfully in open business models, useful ideas will be bountiful. As individuals and companies discover new and exciting technologies, they will receive some degree of IP, which enables them to work with others to commercialize these ideas (via IP markets) or to take these ideas to market themselves. This will provide a wellspring of innovation in companies both large and small. Universities and individual inventors alike will find ready markets for their discoveries and contributions. Consumers will be delighted with new offerings, while companies will get to market faster with higher-value products and services. R&D will provide a compelling return on investment, and more investment will be made in finding new ideas, which can fuel the virtuous cycle once more.

Consider a young firm with the intriguing name of InnoCentive. This company is located outside Boston, but many of its key resources are located around the world. At last count, more than eighty thousand people have used InnoCentive's process to tackle tough problems that companies are trying to solve and win a sizable reward when they succeed in doing so.[17] InnoCentive is putting this resource to work for companies in the chemicals and life sciences industries. Its clients are leading, established firms like Eli Lilly, Procter & Gamble, BASF, DuPont, and Dow Chemical. Even companies as big as these find it wise to seek out ideas and solutions from people located in countries like China, India, Russia, and Israel.

However, there are threats, as well as opportunities, to manage. If you do not deal with intermediate innovation and IP markets effectively, useful external ideas may come to you entangled in a jungle of potentially competing claims. Those ideas may involve IP that could pose a hazard to the unprepared company. If proper steps haven't been taken during the consideration of the IP, owners of the IP (who may or may not be the inventors) may exact a heavy toll on any who dare to tread near their IP. For example, many universities these days use patent royalties as a primary source of funding, effectively privatizing the research that their faculty produces. Anyone using this university output will have to "pay to play." Individual inventors will find it expensive to get a patent and will be unsure of whether that patent is enforceable. They can only be sure that it will cost a lot of money to find out. If inventors attract investors to help with these costs, those investors may have their own agenda for how to make some money from the patent.

Some parties will emerge as aggregators, buying up bits of IP here with other bits there, to erect a thicket of IP. Those businesses that need access to this thicket of IP to practice their technologies will be in a poor negotiating position. They will find themselves at the mercy of these aggregators—who may have had nothing to do with discovering the idea but are legally positioned to profit from any and all uses of it.

BEWARE THE PATENT TROLLS

These predatory IP aggregators are sometimes called "trolls" for their habit of lurking out of sight until an unsuspecting victim comes along. The best known of these trolls is Jerome Lemelson. Lemelson was a prolific inventor (with more than five hundred patents to his name, second only to Thomas Edison in the number of patents he has received). But Lemelson also invented a clever—and legal—way to play the patent game. He pioneered the ability to file a patent claim privately with the patent office and then revised his claims periodically before the patent issued. These revisions served to keep the claim private, while allowing him to update the claims of the patent to make them more commercially relevant and valuable. In many cases, the eventual claims that issued with the patent bore little resemblance to the initial claims made. Lemelson received hundreds of patents that claimed inventions in commercially useful areas such as television, computers, and other leading-edge industries. Hundreds of companies have had to pay the toll that Lemelson exacted.[18]

INTERMEDIATE MARKETS ARE COMING: IS YOUR BUSINESS MODEL READY?

Open Innovation exploits the new division of innovation labor. This division shows up in intermediate markets for innovation and intellectual property, where ideas and their legal protection can be bought or sold. These markets will dramatically change the innovation environment for all of us. Like other markets, these markets can be managed once they are understood. To take full advantage of them, you must open up your business model, even as you take steps to guard against the risks.

Managing Open Innovation in a world of intermediate markets for ideas requires the construction and support of a rich internal innovation network. This network in turn must connect the firm to an incredibly large, diverse external innovation community. The business model and structure of management itself will have to change for the mind-set of open business models to become part of corporate management. A few companies have made this transition already. For most others it will require more substantial changes to the business model and to the structure of innovation management processes.

While this will entail an enormous amount of work, the results will make it well worth doing. Shorter product life cycles and accelerating costs spell doom for the older, closed model of innovation. Only by accessing more ideas, and by using these ideas in more new products and services, can organizations keep pace and achieve an economically viable model of innovation. And only through innovation can companies deliver consistent organic growth to their shareholders, their customers, and their employees.

Companies' business models will shape the world in which they compete in the twenty-first century. There are alert, astute companies today creating open business models that will help them innovate through a global marketplace of ideas, both as suppliers and as customers. Companies that ignore the opportunities in this global marketplace are leaving money on the table. Companies that are oblivious to the dangers in accessing ideas and IP in the global marketplace are making themselves easy marks for the sharp but legal practices of others.

2

The Path to
Open Innovation

In my previous book, *Open Innovation*, I argued that companies should organize their innovation processes to become more open to external knowledge and ideas.[1] I also suggested that companies let more of their internal ideas and knowledge flow to the outside when those ideas and knowledge are not being used within the company. It turns out that the path to Open Innovation is not a simple, straight one; rather, it is filled with twists and turns, with unexpected surprises along the way. This chapter will discuss the path that companies must walk to open up their innovation processes. It will also discuss some examples of companies that have faced these challenges and have strengthened their businesses by opening up.

I have organized this chapter by starting with the issues companies must face in bringing outside ideas into the company. This can be thought of as an "outside-in" process. Then I will consider the "inside-out" process, where companies must overcome barriers to taking unused internal ideas to the outside.

USING EXTERNAL INNOVATION
INSIDE THE COMPANY

A key idea in the book *Open Innovation* is the notion that "not all the smart people work for you." Instead, useful knowledge is increasingly dispersed in companies of all sizes, in many parts of the world. More recent data that has come out since the book was published continues to suggest a more level playing field for industrial innovation activity. Data on the United States from the National Science Foundation, in table 2-1, shows that small firms (defined here to be those firms with fewer than a thousand employees) continue to increase their share of the total industrial R&D spending, amounting to almost 25 percent of total industry spending in 2001. Large firms (defined here as firms with more than twenty-five thousand employees) saw their collective share of industrial R&D fall to less than 40 percent of total industry spending in that year. As the table shows, this is an increase even from the prior survey, reporting spending activity in 1999, which was the data I used in *Open Innovation*.

These data and other data, such as the growth of employment in small enterprises relative to employment in large firms, all combine

TABLE 2-1

U.S. industrial R&D spending by size of enterprise, selected years, 1981–2001

	1981 (%)	1989 (%)	1999 (%)	2001 (%)
Fewer than 1,000 employees	4.4	9.2	22.5	24.7
1,000–4,999 employees	6.1	7.6	13.6	13.5
5,000–9,999 employees	5.8	5.5	9.0	8.8
10,000–24,999 employees	13.1	10.0	13.6	13.6
25,000+ employees	70.6	67.7	41.3	39.4

Source: For 2001: National Science Foundation, Division of Science Resources Statistics, "Research and Development in Industry: Science & Engineering Indicators–2004," table 4-5. For prior years: Henry Chesbrough, *Open Innovation* (Boston: Harvard Business School Press, 2003), 48, citing earlier National Science Foundation reports from its Resource Studies unit.

to suggest that the playing field for innovation is becoming more level. Put differently, there are fewer economies of scale in R&D than there were a generation ago.[2] This more level playing field has powerful implications for the organization of innovation. In a more distributed environment, where organizations of every size have potentially valuable technologies, firms would do well to make extensive use of external technologies.

So far, so good. But this is not an easy change for companies to make. One challenge every company must face in becoming more open is the internal resistance to external innovations and technologies within the company. This internal resistance has long gone by the name of the "not invented here" (NIH) syndrome. This NIH syndrome is partly based on an attitude of xenophobia: we can't trust it because it is not from us and is, therefore, different from us. But there are more rational components that might induce internal employees to reject external technologies as well.

Managing Risk in Internal R&D Projects

One such component is the need to manage risk in executing R&D projects, especially when the cycle time to complete a project is accelerating. When cycle times accelerate in a project, there is less time to evaluate external technologies and incorporate them into it. More subtly, when projects are moving fast, project leaders seek to minimize the risk of unexpected outcomes in the project. Internally sourced technologies already pose enough risk to the likelihood that a project will meet its scheduled ship date. Externally sourced technologies, coming from a much wider variety of sources about which much less is known (when compared to internally generated technologies), may greatly increase the perceived risk to the project. So an externally sourced technology may have the same average estimated time to complete, but it may have a wider range or variation in that estimated time relative to an internally created technology.

A more subtle challenge is the impact on the internal staff's subsequent actions if and when externally sourced technologies are used and prove to be highly effective. In this instance, the overall project's success may be enhanced by the inclusion of externally sourced technology. But the top managers in the firm might infer from this experience that the firm doesn't need quite so many internal R&D staff to accomplish the *next* project, that the next project ought to rely more on external technology as well. In this case, the short-term success of the project might be to the long-term detriment of internal R&D staffing levels and internal research funding. So the project team confronts a risky situation in which it bears full responsibility if the use of external technology fails, yet the internal R&D organization may bear other long-term costs if the use of external technology succeeds.

Overcoming the "Not Invented Here" Syndrome

How have firms overcome these challenges? Well, it probably helps to be a young, rapidly growing firm in a fast-changing industry. This context of rapid growth means that there is no risk to internal employees being displaced from the incorporation of external technology, because the firm has chosen not to build such a staff in the first place. Intel's founders, for example, were constantly reminded of their earlier experience with unconnected research labs at their previous employer, Fairchild. They consciously chose to grow as much as possible with as little investment in research as possible.[3] More recently, Dell innovates extensively with the inventions of others, rather than develop a large internal R&D laboratory. Similarly, Cisco Systems has developed and refined a model of "A&D" (acquire and develop) instead of the traditional R&D. These firms have grown up while relying on external technology, so there is little or no risk for their internal R&D organizations in exploiting external technology.

What about older and more traditional firms? How have they overcome these challenges? An interesting case here is Air Products, an industrial supplier of gases used by manufacturing companies. Air Products overcame these challenges by fundamentally changing its business model. In the old days, Air Products got paid by the railcar for the industrial gases it delivered to its customers. Once the cargo was received, Air Products got paid, and how the customers used these gases was not of concern to the company.

More recently, Air Products altered its business model to take responsibility for the provision of industrial gases directly on the manufacturing floor. If there were leaks, or if the gases were used inefficiently, Air Products would receive less money from its customers. Now it had a strong business incentive to apply its considerable know-how about the proper handling of gases to increase the efficiency of its customers' use of its products. And any ideas that could increase the use of its gases would increase Air Products' payments—regardless of whether those ideas came from Air Products itself, or from others.

Dealing with Context in Overcoming "Not Invented Here" Syndrome

The context in which an Open Innovation approach is presented within the company also affects the level of internal resistance. Large, mature firms can only adopt a more externally oriented technology strategy after the internally oriented strategy is widely seen as having failed. It usually requires a significant downsizing of R&D staff for this failure to be perceived. Once these layoffs have occurred, the (remaining) internal staff may realize that unless a more successful approach to R&D is found, their own positions may be at risk. In this instance, the reference point for the internal R&D staff has been shifted by the recent layoffs. Because recent experience suggests that the firm is unlikely to return to a largely internal R&D approach, there is far less to lose from embracing external technology.

This context arose in two important cases cited by *Open Innovation*: IBM and P&G. In IBM's case, the company operated with a highly vertically integrated and inwardly focused innovation model since the inception of its System 360. IBM's shift toward a far more open, less vertically integrated approach came from the arrival of CEO Lou Gerstner. However, immediately before Gerstner became CEO, IBM reported what was at that time the largest quarterly loss in U.S. business history, and IBM made the first major layoffs in its corporate history. This dramatically shifted the previous culture of internal focus toward innovation, since many of those laid off were in the R&D organization. When IBM began to adopt a more open approach, it did so at a time when the organization had recognized that the former status quo was no longer sustainable.[4]

Procter & Gamble's embrace of Open Innovation also was immediately preceded by a significant layoff in its R&D organization (though this was a far less severe layoff than the one at IBM). P&G had embarked on a growth campaign in 1990 to double its revenue by 2000. When 2000 came, the organization had fallen short of its goal, and many spoke of the "$10 billion growth gap" for the company. P&G cut expenses significantly and laid off a large number of people. After these cuts, P&G consciously told its R&D staff that its shift to what it called Connect and Develop would not lead to any further reductions in staff. Instead, the shift in innovation models was positioned to enable P&G to generate more innovation with the (recently reduced) R&D resources on hand.[5] This shift would have been received very differently by P&G's internal organization had Connect and Develop been launched *before* the layoffs of P&G's R&D staff.

OFFERING UNUSED TECHNOLOGIES
OUTSIDE THE FIRM

Another aspect of becoming more open comes from looking at the innovation process from the other end, where companies choose to deploy certain internal technologies and commercialize them, while leaving a larger set of internal ideas and technologies unutilized. One of the dirty little secrets of more traditional innovation processes is that many of the ideas and technologies developed within the company never get used either inside or outside the company.

Why bother with these unutilized ideas? There are a number of reasons. First, unused ideas are a waste of corporate resources. Second, unused ideas are demoralizing for the staff that created them. Third, unused ideas clutter and congest your innovation system, slowing it down. Fourth, releasing unused ideas outside will generate new knowledge about market or technical opportunities—which would never emerge if these ideas were kept bottled up inside the firm. And last, if ideas get bottled up too long, they may find another, unplanned exit of their own. They might leak out to another firm, or an internal group of people might choose to take them out on their own.

Reasons for Unused Ideas and Technology

Unused ideas abound in many companies. When Procter & Gamble surveyed all of the patents it owned, it determined that about 10 percent of them were in active use in at least one P&G business, and that many of the remaining 90 percent of patents had no business value of any kind to P&G.[6] Dow Chemical went through an extensive analysis of its patent portfolio starting in 1993.[7] In that year, about 19 percent of Dow's patents were in use in one of its businesses, while a further 33 percent had some potential defensive use, or future business use. The remaining patents were either being licensed to others (23 percent) or were simply not being used in any discernable way (25 percent). In the typical pharmaceutical development process, a company must screen hundreds or even thousands of patented compounds to find a single compound that makes it through the process and gets to market.[8] From a naive perspective, it seems very wasteful to create and develop a large number of technologies and then only use a small fraction of them in any way, shape, or form.

The Connection to the Business Model. The reason for such low utilization levels is that many firms consciously keep the research portion of their R&D process only loosely coupled to their business model.[9] As I argued in chapter 8 of *Open Innovation*, most companies have a very decentralized process for determining what projects research staff work on and what invention discoveries get made, and a similarly decentralized process for deciding whether to patent these invention discoveries. Many R&D departments recruit staff by promising prospective employees extensive freedom in what research they do and often compete against universities for hiring these people. So these organizations consciously limit the coupling of research output to any business model.

Further, R&D managers often use the number of patents generated by an R&D researcher or an R&D organization as a metric to judge the productivity of that person or organization. Similarly, some R&D

organizations count the number of publications generated by their R&D staff as another measure of productivity.[10] Unsurprisingly, when an organization rewards the quantity of patents or papers produced, the R&D organization responds by generating a large number of patents or papers, with little regard as to their eventual business relevance.

Budgetary Disconnects in the Business Model. To carry this point further, there may be a budgetary disconnect between a research and development group on the one hand, and a business unit on the other. To see this, examine figure 2-1.

In this figure, R&D produces research results and operates as a cost center. This is usually how such organizations are funded, since they do not sell their output directly, and since it is hard to estimate how much money a particular R&D project will need to be successful. Instead, companies determine an amount of funding they can sustain over time that can be dedicated to R&D tasks. The R&D unit manager must in turn decide how many projects to support with the funds she has that period. It is bad for her to exceed her budget, since the organization may not be able to sustain the additional expenses. It is also bad for her to come in much under the budget that year, because that may suggest that next year's budget can be reduced as well. So the manager tries to develop as many projects as she can, subject to the budget constraint.

The internal business unit customer, by contrast, is typically managed on a profit-and-loss (P&L) basis. The business unit typically does sell its output to customers, and giving each business unit its

FIGURE 2-1

A model of budgetary disconnection between R&D and the business unit

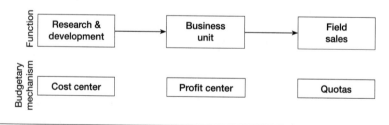

own P&L enables that business manager to make the best use of his resources to maximize profits for the business. That manager wants to buy low and sell high. So the business unit manager wants any R&D project coming from his internal "supplier" to be as fully developed as possible. This reduces any additional costs the manager must incur before using the technology in the business. It also reduces any risk to that business's profitability during that period.

The stage is now set for the budgetary disconnect between the two functions. The R&D manager wants to push out the project as soon as the publications and patents have been generated. Further development within the R&D budget crowds out other, newer projects that have greater potential for generating new patents and publications. So the R&D manager's incentives are to transfer the project sooner rather than later to the business unit. Meanwhile, the business unit manager's incentives are to wait as long as possible before taking over the further funding of the R&D project on his P&L.

The resolution of this budgetary disconnect is to place a buffer between the R&D operation and the business unit, as shown in figure 2-2. This buffer provides temporary storage for the R&D project, until the time when the business unit is ready to invest in its further application within the business. This lets the R&D manager move on to her next project, without requiring the business unit manager to commit to further funding on his P&L until he judges it to be beneficial.

While this solves the local problem of each manager, from a system viewpoint it causes many R&D projects to pile up in the buffer.

FIGURE 2-2

Decoupling R&D from the business unit

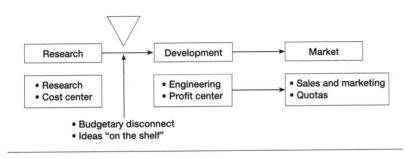

These projects are often termed "on the shelf," because they are no longer being actively pursued by the R&D organization, nor are they actually being used by the business unit.

How can this challenge be overcome? Some research organizations obtain a significant percentage of their funds from research contracts with their internal business units. These contracts tend to be fairly specific and near term, and the research outputs of these contracts are likely to be used by the business units, which pay directly for them. But other funds for those same research organizations come from a corporate allocation of funds (which is generated from a "tax" on all of the businesses within the firm). These corporate funds are not tied to any specific business unit objective and are allocated to longer-term projects whose output may benefit multiple businesses. Still other research funds come from government research contracts. These funds tend to be reviewed by academic peers and may therefore have little or no relevance to any business unit activity within the firm.

In IBM's R&D organization, the company tries to establish research contracts between IBM business units and IBM research teams to foster a stronger connection between the goals of the business and the output of the researchers. While this has helped increase business utilization of IBM research results, it is not a complete solution to the problem. For one, the business units have a clear sense of their needs in the next one to two years but have much more limited insight into their needs over a longer period. For another, many research outputs have uses in multiple areas, some of which may not be addressed by any single business unit—at least not yet.

Barriers to Greater External Use of Unused Internal Ideas

The foregoing analysis suggests that R&D processes are only loosely coupled to the business models of firms, resulting in a substantial number of technologies that are un- or underutilized within those businesses. This prompts a second question: what prevents the business from enabling others to use those underperforming technologies in their own respective businesses?

Biases of the Current Business Model. One issue may be that companies think that if they cannot find a profitable use for their technology, no one else will either. If companies were truly objective in their assessments, that might be so. But the internal view of the technology's potential is likely biased by the business model of the company. An external view of the technology's value may be more unbiased (if less informed, at least initially) than the internal view. But by itself, this analysis would suggest a potentially substantial market for underutilized technologies. After all, when buyers have higher valuations of projects than sellers, those sellers can find a mutually beneficial transaction that sells those projects to the party with the higher valuation.

A second concern may be adverse selection. Buyers may worry that the sellers of unused technologies will only offer the "bad" ones. This worry presumes that both parties are rational and unbiased, in which case the seller (who has more information sooner) will inevitably have an advantage over the buyer. But the dominant logic of a company's business model would actually suggest a countervailing consideration. While selling companies have significant prior information on a technology project, that information will be interpreted within the context of the company's business model. If the buyer possesses or can identify a very different business model for that technology, the buyer's evaluation of the project may differ greatly from that of the seller. Put differently, the buyer may see an opportunity that is not visible to the seller, because the buyer has a different business model in mind than the seller.

To give an example here, consider the experience of Xerox PARC with its many technology spin-off projects.[11] I have identified thirty-five projects that left Xerox after funding for the work had ended. Xerox judged that there was little or no additional value to be gained from continuing this work. In thirty of the thirty-five projects, Xerox even gave a license for the technology to the departing spin-off, so most of these separations were consciously managed departures, not inadvertent oversights. In twenty-four of the thirty-five projects, there was little business success after separation. But for eleven of the projects, each of which developed under a very different business model from that of Xerox, there turned out to be substantial value. The collective market value of the companies that emerged from these eleven projects turned out to exceed the total market value of Xerox by a factor of two. I interpret this data to mean that Xerox's estimates of the value of these projects were biased by its business model. In direct interviews, many of the participants in these events acknowledged that they never dreamed that some of these projects would become so valuable.

Not Sold Here. Other barriers to greater utilization of unused technology may lurk as well. There may be a behavioral analogue to the NIH syndrome that sits within the business units, which I term the "not sold here" (NSH) virus. NSH is a virus that argues that, if we don't sell it, no one should. It is rooted in the surface perception that, if our organization cannot find sufficient value in the technology, it is highly unlikely that anyone else can either. At a deeper level, however, the NSH virus seeks to forestall competition with outside entities for accessing internal technology. Most business units enjoy a monopsony position relative to their R&D unit suppliers. Because they have exclusive rights to the technology, they can defer costs and delay commitments to the technology without incurring any penalty to their unit for waiting longer before using that technology.

Enabling greater external use of unused technologies alters the business unit's calculation. Let's assume that a business unit chooses not to incorporate a technology, and that the company now has a process that allows others the chance to do so. The business unit faces a previously latent cost to waiting: if it doesn't use the technology itself, it might "lose" that technology to an external organization. Under Open Innovation, internal business units have some defined interval of time during which they can claim the technology. After that interval expires, the technology is then made available to other firms.[12] Depending on who that external firm is, the internal business unit may even have to compete against that technology in the market. Worse (from the business unit's perspective), the external use of the technology might reveal previously unrealized value from the technology, leaving the business unit in the awkward position of explaining why it failed to use this apparently valuable technology. Another challenge presents itself: if the technology is licensed externally, the corporation may "win" through additional licensing revenue, but the business unit may "lose" through additional competition in its market. Many companies conclude that inside-out mechanisms are just a distraction, with little strategic value. Events, however, have not supported that conclusion, as some companies are making real money from this "distraction."[13]

Other Aspects of the Inside-Out
Equation to Consider

Overcoming the NSH virus will require other actions as well. Particular attention will need to be paid to the people involved in the innovation process and to the continued investment needed to support innovation.

Aligning Incentives for Greater External Use of Ideas. One way to overcome NSH is to develop mechanisms that the company can employ to align incentives within the business unit to more closely approximate those of the overall firm. GE and IBM, for example, both make many technologies available for external use. They manage the internal business resistance to this by sharing any licensing revenues or equity participation from a technology with the business unit associated with the technology. So the business unit P&L not only bears the risk of competing with the technology in the market (thus negatively impacting the P&L of the unit), but it also receives credit for licensing revenue or equity upside from the technology on its P&L (thus boosting the revenue and profit of the P&L of the unit).

The Human Cost of Unused Ideas and Knowledge. There is a further, more human business rationale for enabling greater external use of unused technologies. Companies in which NSH is dominant likely frustrate their R&D staff, because many of the ideas these people work on are never deployed in the market. It is reportedly quite common for a pharmaceutical researcher to never see one of her projects ship to market, even over a thirty-year career, because the attrition rate of compounds is so high. This is an enormous waste of human talent and must take a toll on any person's initiative. Companies that can overcome NSH begin to allow other pathways for internal ideas to get to market. These other pathways allow the market to provide feedback on those ideas. This helps researchers see their ideas in action in the wider world, even if those ideas do not make it into the company's own products. This also provides new sources of feedback for the researcher on how to improve on those ideas. Some of those improvements might one day make it into the company's own products. Thus, it can be an act of enlightened self-interest for companies to let more unused ideas flow outside the company.

The Continuing Importance of Internal R&D. If leveraging external technology is so helpful, and if allowing unused ideas to go outside can create such positive outcomes, one might reasonably ask, why bother with internal R&D in the first place? While there have been some very helpful academic articles written to answer this question, the basic answer is quite simple.[14] You cannot be an informed consumer of external ideas and technology if you don't have some very sharp people working in your own organization. Not all the smart people work for you, but you still need your own smart people to identify, recognize, and leverage the work of others outside your company.

There are additional benefits to internal R&D in the Open Innovation context. One benefit is that your smart people can fill in the gaps in what others are doing, to finish the solution that your company needs to address an important problem for your customers. Another, more subtle benefit is that your smart people can forge the architectures and systems knowledge that can organize and direct the work of others. This will be discussed further in the type 6 business model in chapter 5.

LEVERAGING OPEN INNOVATION
IN A START-UP

In large companies, Open Innovation relates to buying or selling technologies and their associated IP as part of the business model. In smaller companies, there usually is less IP to buy or sell. While smaller companies can still buy or sell IP, for them Open Innovation more frequently involves collaborations and sharing of technology and IP with other parties as part of the business model. The next two case studies examine one particular problem that is especially acute for the business models of small firms: how to attract customers, capital, and employees while preserving your most sensitive information.

GO: Can You Be Too Open?

Innovation is probably even more important to manage well in a small company. But small companies are faced with a dilemma: on the one hand, they must protect their ideas and technologies as much as they can so that bigger companies with more resources don't steal their ideas; on the other hand, they need to raise capital, hire employees, and attract customers in order to survive and grow. This latter influence requires them to disclose a great deal about their ideas, technologies, and plans before they have taken steps to protect their ideas. It is a challenge to strike the right balance.

One software start-up company that (with the benefit of hindsight) erred on the side of too much disclosure was GO Corporation. This company developed an operating system for pen-based personal computer products called PenPoint. It was founded in 1987 by Jerrold Kaplan, who had been the chief technologist at Lotus, and was funded by Kleiner Perkins Caufield & Byers, a premier venture capital firm.

While GO was an initially promising venture, its product never became widely established. The company was eventually acquired by AT&T and was subsequently shut down. While most failed software companies are consigned to obscurity, Kaplan wrote and published a book about GO's experience, Startup.[15] This book provides a fascinating inside perspective on GO's struggle to manage its dilemma. GO's story is varied and complex. With the benefit of hindsight, though, it seems clear that GO chose a business model that required it to be too open to the wrong people.

GO faced the problem of many start-up software companies, the need to attract outside firms (customers, suppliers, and third-party applications software vendors) to support its technology. In particular, GO needed to recruit software developers to create software applications using its PenPoint operating system. Since Microsoft was the largest applications software developer for both the Windows and Macintosh operating systems, GO met extensively with Microsoft to encourage Microsoft to develop applications for GO's Pen-Point operating system. To protect GO's business, the two companies met after Microsoft executed a nondisclosure agreement with GO.[16]

Earning Microsoft's support for the PenPoint operating system had vital strategic ramifications. If Microsoft were to develop applications for PenPoint, other software developers would then be more likely to develop for PenPoint as well. Potential customers would also have been more likely to purchase PenPoint, since they could have expected more applications to emerge on the PenPoint platform.

GO could not earn Microsoft's support, though, without disclosing an extensive amount of its proprietary information. Bill Gates himself spent an entire day at GO, along with a technical engineer, reviewing GO's technology, product strategy, and business plans in detail. Gates's engineering colleague returned later to hold additional meetings with GO personnel. However, instead of building applications for GO's PenPoint operating system, Microsoft elected instead to launch its own competing PenWindows operating system six months later. Much to GO's surprise, the PenWindows effort was headed by the same Microsoft engineer who visited GO on those previous occasions!

GO's mistake was fatal, for Microsoft was not only an applications developer. First and foremost, Microsoft's business model made it focus on leading the industry in operating systems. Whatever business benefit Microsoft might have obtained from selling applications to a new market segment of pen-based computers was dwarfed by the competitive threat to its dominance of the PC operating system market. Microsoft's entry into pen-based computing "froze" many software developers who might otherwise have supported GO and caused potential customers to wait and see what PenWindows would be like. It cost GO tremendous momentum and eventually sank the company.

The question arises, why didn't GO sue Microsoft for misappropriating its ideas? It thought hard about doing so but had to face the realities of its IP protection and its limited resources. Software copyright protection is limited to the actual code used in a product and provides no protection for the overall concept and algorithms.

Microsoft presumably did not use GO's own code in its product, so there may have been no actual legal infringement.[17]

Collabra: A Start-up Business Model with Strategic Openness

Another software company that started a few years later, Collabra, arguably did a better job of figuring out how open to be and with whom to be open. This company was founded in April 1993 and was acquired two and a half years later for $107 million by Netscape in October 1995. Most start-up companies like this one are hard to study in retrospect because their experiences are usually lost after integration with a larger company. However, through the cooperation of Eric Hahn, the founding CEO of Collabra, I was given access to all of Collabra's nondisclosure agreements (NDAs) from the time of the company's founding to the time of its acquisition.[18] This allowed me the rare opportunity to reconstruct the company's approach to sharing and protecting its ideas.

The company developed a software product that allowed multiple users to collaborate jointly on the creation and editing of documents. While based on a different technical approach, the product competed directly against Lotus Notes and provided a similar functionality to Notes (though Notes was a larger, more sophisticated program).

In all, Collabra executives executed 195 NDAs, from Collabra's initial formation in April 1993 to its acquisition by Netscape in October 1995.[19] Figure 2-3 shows the number of NDAs signed in relation to major announcements by the company.

Although Collabra was formally founded in April 1993, the founder actually executed some NDAs before the formal incorporation of the company. During the twelve months after its incorporation, the company was in stealth mode, when it deliberately kept a low public profile. As the graph shows, though, this stealth period was also when the company signed the greatest number of NDAs. The company's first product was unveiled in March 1994 and shipped in July of that same year. There were relatively few NDAs signed after that point, up until the time the company was acquired by Netscape in October 1995.

FIGURE 2-3

Collabra's NDA agreement history

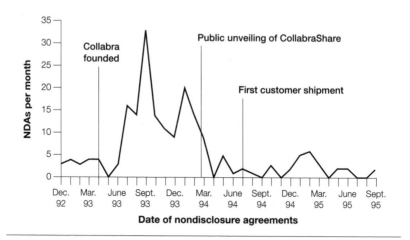

Date of nondisclosure agreements

Results that tabulate these NDAs for each of these recipient groups are shown in table 2-2.

Collabra was actually very open with its customers. As the table shows, the leading recipients of NDAs were customers, who executed 26 percent of the NDAs in the dataset. The company also was very open with third-party software developers, who executed 22 percent of the NDAs. Suppliers signed 16 percent of the NDAs. Thus, almost two-thirds of the disclosures in the dataset went to outside organizations that had to make complementary investments in order to support Collabra's product. Collabra knew that its own value would be enhanced if customers bought its product and if third parties made their own investments to develop products that enhanced Collabra's value. Collabra also knew that it should be careful in sharing its ideas, even with friends and previous coworkers. So the company made these people also sign NDAs before telling them what it was doing. Informal friends and associates were the third-largest recipient group, receiving 39, or 20 percent, of all of the disclosures. Employees received 7 percent of the disclosures made, while the general group of press/analysts and competitors received 9 percent of the disclosures made.[20]

TABLE 2-2

Frequency of Collabra disclosure by type of recipient

Recipient type	Number of NDAs	Percent of total
Customers	50	26
Suppliers	32	16
Third-party software developers	43	22
Friends, associates	39	20
Press, analysts, competitors	17	9
Investors	0	0
Employees, contractors	14	7
	195	100%

Source: Collabra nondisclosure agreements, 1992–1995.

A key issue for Collabra was how open to be with its direct competitor, Lotus, and its potential competitors, in particular, Microsoft. Collabra chose not to disclose to Lotus at all. However, Collabra did meet often with Lotus Notes customers and resellers and disclosed its plans to them under an NDA.

Microsoft was more difficult to manage. Collabra thought hard about whether and how to approach Microsoft. One of Collabra's marketing managers had previously worked at GO Corporation and knew the history of that relationship quite vividly. Collabra obviously didn't want to share GO's fate and didn't want Microsoft to compete directly against it.

The Enemy of My Enemy Is My Ally. Given that Collabra was competing with Lotus Notes and wanted to compete successfully against this bigger company, though, Collabra knew that it needed some allies. For these reasons, it felt that it needed to find a way to ally with Microsoft. Collabra also knew that, in the short term, Microsoft needed it as well, in order to compete against Notes. Microsoft's own efforts in this area were lagging. Collabra's products, when teamed with those of Microsoft, would boost Microsoft's credibility against Lotus, while Microsoft's cooperation with Collabra would give the start-up firm significant publicity and credibility. An alliance with Collabra would buy time for Microsoft to bring its own offering to market.

Collabra was well aware of the risks this entailed and fully assumed that Microsoft would abandon it down the road. Collabra also was careful to limit disclosure of its technology both to allies and competitors, and generally preferred to work with lower-level representatives of companies, instead of the senior leadership. Microsoft's sales representatives, for example, were very helpful to Collabra and far from Microsoft headquarters in Redmond, Washington. The risks of being supplanted by Microsoft depended in part on how quickly Collabra got its product to market relative to how quickly Microsoft got any competing products to market.

Ultimately, Collabra made a pragmatic decision to take the risk. Microsoft signed an NDA with Collabra on December 16, 1993, three months before Collabra's first public unveiling of its product at an industry conference. Collabra subsequently participated in marketing seminars with Microsoft and benefited from the publicity this alliance provided. By the time its first product, CollabraShare, shipped in July 1994, Collabra had achieved a significant presence in its marketplace by virtue of its publicity efforts, including its joint marketing with Microsoft.

Venture Capital Investors: Unwilling to Sign NDAs. One surprising result is that no investors signed any of Collabra's 195 NDAs, even though the company held some of its earliest (and presumably most sensitive) discussions with potential outside investors such as venture capitalists (VCs). In interviews with former Collabra executives, I probed this point. I was told that VCs as a rule refuse to sign such documents. They claim that in order to develop investment syndicates with other VC firms to make investments in start-up firms, it is occasionally desirable to share some information about promising new investment opportunities. The claim is that NDAs would impose undesirable restrictions on the ability of such investors to create syndicates that might invest in the firm.

Discussions with venture capital partners confirmed that this is their practice. VCs look to make high returns on their investments while also seeking to syndicate investments to spread the risk from any individual investment. Syndication also brings another independent investor into the deal to judge the risks, returns, and value of the investment. It is common practice, for example, to have a new investor coming into a new round of financing set the price for the stock for that round. To achieve this syndication, VCs freely trade information with one another about new ventures under consideration. They therefore refuse to sign NDAs.[21]

Maintaining Leverage While Being Open. While its VCs never did sign an NDA, Collabra was not shut down or forced to join another company. The company began to pursue an initial public offering and also explored the possibility of being acquired. Eventually, Collabra was able to interest three different companies in buying the company. This required the sharing of a great deal more information about the company's business. In addition to Netscape, which eventually bought Collabra, Novell and Microsoft also were interested. This helped the small company maintain some negotiating leverage over these much larger firms. In October 1995, Netscape bought Collabra for roughly $100 million.

Lessons Learned from GO and Collabra

Small firms like GO and Collabra offer some important lessons for how to manage Open Innovation. One lesson is that you can be too open. If you share the wrong information with the wrong people, it can kill your company. It is a mistake for small companies to rely entirely on formal legal protections, backed up by the court system, to enforce their IP rights. A small company's scarce resources limit the amount of protection it is likely to receive. While small companies should obtain as much protection as they can afford, there is no substitute for a good business model to protect IP. GO chose a business model that placed the company in direct competition with Microsoft in operating system software, but one that required Microsoft support in applications software in order to succeed.

Collabra chose a business model that competed directly against Lotus and harnessed Lotus's competitors (like Microsoft) in order to succeed. In Collabra's business model, its competitor's competitor was a reliable ally, at least in the short term. Collabra achieved some temporary protection of its ideas with Microsoft because it knew Microsoft would benefit from working with the smaller company, not because of the strength of Collabra's IP. And Collabra always maintained an outside option by courting other potential strategic partners—whether those other partners were Novell, IBM, or Netscape—if Microsoft didn't respect its IP.

MANAGING OPEN INNOVATION: IT TAKES AN OPEN BUSINESS MODEL

Whether you are in a large organization or a small one, chances are you need to open up your innovation processes. But to do this effectively, you must connect your business model to your innovation process. Companies that are large typically enjoy strong business models. However, they face challenges and risks that small companies do not. It is hard for large companies with successful business models to change them to exploit Open Innovation opportunities. Small companies, on the other hand, lack the strong business model and resources to enable them to exploit the opportunities of Open Innovation without fear of being copied by a larger foe. IP protection is one of the many tools needed in their business model to achieve success.

In the next chapter, we will look more carefully at the external environment surrounding the company's business model. Before doing that, however, we must consider a frequent objection that is raised about the idea of managing Open Innovation in the context of your business model: what about open source software development?

OPEN SOURCE: A SUCCESSFUL TECHNOLOGY WITHOUT AN APPARENT BUSINESS MODEL

One of the central tenets of my previous book, *Open Innovation*, is that business models are essential to unlocking latent value from a technology. In the introduction, the book asserts: "There is no inherent value in a technology per se. The value is determined instead by the business model used to bring it to market. The same technology taken to market through two different business models will yield different amounts of value. An inferior technology with a better business model will often trump a better technology commercialized through an inferior business model."

Open source software development seems to challenge this claim. By construction, open source software is created without any one firm owning the ability to exclude others from using technology, provided that these other firms observe the open source requirements. Enhancements to the code are available to everyone on an equal basis.

Is open source's success simply an exception to the general rule, is this success due to a business model of a different kind, or is there something fundamentally wrong with the claims of *Open Innovation* regarding the importance of business models? What I hope to show now is that the evolution of the open source software movement is being propelled by the emergence of clear and distinct business models built around it. I think of them as "open source business models."

Open Source Software: A (Very) Brief Introduction

There have been numerous studies of the open source software community, ranging from enthusiastic proclamations of its benefits (e.g., Eric Raymond, *The Cathedral and the Bazaar*), to condemnations of it as pernicious to innovation (such as the comments by Steve Ballmer, CEO of Microsoft, likening Linux to a "cancer").[22] More scholarly examinations of the open source software approach can be found in "The Simple Economics of Open Source," "Guarding the Commons," and "How Open Is Open Enough?"[23] Listings of open source research resources are abundant online.[24]

And this wealth of study takes no account of the enormous online literature that discusses many elements of open source. This ranges from blogs, to communal places like Slashdot (www.slashdot.org), to repositories of open source software code, such as SourceForge.net.

As a result, we now know a great deal about how open source software development works. It is a collaborative, community model of development, based on a process that does not allow any contributor to exert a proprietary claim to intellectual property on any portion of the code being developed within the open source framework. (However, the technical legal status of open source software is actually varied and complex, as different projects employ different licensing arrangements, which vary in the rights those arrangements convey to developers to use their contributions in other, proprietary software.)

What You Don't Read About:
Open Source Business Models

One doesn't read much about business models in open source software development. There are strong social norms and legal protections that have been crafted to discourage people from profiteering on the work of their peers. There are even frequent postings on well-frequented Web sites that identify cases where the norms of the group appear to be violated (though there have been few, if any, legal sanctions against violators).

However, occasional crises reveal parties that have developed business models to profit from the adoption of open source software. One crisis of this kind was the threat by the Santa Cruz Operation (SCO) to enforce its alleged IP rights (contained in a version of Unix it purchased from Novell) over the code being widely circulated in the Linux community. It subsequently sued IBM for $1 billion for allegedly devaluing SCO's Unix license. Separately, SCO sued users AutoZone and Daimler Chrysler for using Linux without a license to SCO's Unix.[25] While the open source community was very upset over these lawsuits, a very different response came from a group of companies that included Intel, IBM, Hewlett-Packard, Novell, and Red Hat. These companies banded together to pool resources into a fund to indemnify customers of open source software for legal expenses they might incur in defending themselves from a lawsuit, should they choose to use open source software. Separately, IBM appears to have taken the lead role in defending against the SCO suit. Without demeaning IBM's sincerity in any way, committing substantial resources in this manner is a highly reliable indicator that IBM's business model, and the business model of the other companies just mentioned, benefits significantly from the adoption of open source. Indeed, IBM has publicly stated its strong support of Linux and devotes more internal software personnel to supporting Linux development than any other single organization in the world. Other important IT companies like Sun Microsystems also have positioned themselves to profit from open source.[26] Even longtime Microsoft allies Intel and Dell have active programs supporting open source.

How can a company create a business model to profit from open source software? There are a number of open source business models. To rank them from lower to higher value added, these models include:

- Selling installation, service, and support with the software

- Versioning the software, with the free version as an entry-level offering and other, more advanced versions as value-added offerings[27]

- Integrating the software with other parts of the customer's IT infrastructure

- Providing proprietary complements to open source software (these increase in value as the cost of the open source code falls; one version of this strategy is to create a creative commons and then build proprietary products or services on top of the commons)[28]

Companies like Red Hat illustrate the first of these business models. While Red Hat sells the Linux operating system like many other companies, it has developed a variety of tools to make Linux easy to install and operate on a variety of computers. So Red Hat doesn't make much money on Linux but captures reasonable profits from the associated tools it has developed.

The versioning business model is exemplified by the open source product MySQL. The entry version is free; the more advanced versions, which provide greater functionality, are more expensive. This model uses open source software to provide the customer with a lower cost of initial acquisition while allowing the company to then sell that customer more proprietary, enhanced products later. Other strategies offer the customer the ability to perform user modifications of the software without any penalty, transferring those costs to the customer. In this scenario, the customer achieves greater customization of the product to his needs, without making the software company develop the custom portion of the code.[29]

IBM is one example of a company that makes money off of open source software like Linux and Java by using those products to help companies integrate them with other parts of IT infrastructure. A good portion of IBM's work in these tools, for example, creates device drivers that help connect other IBM hardware and software together.

Then there are more subtle business models that have emerged in the creative commons arena. One example of such a model is when a company voluntarily chooses to donate portions of its intellectual property to a commons, so that the company and others can practice their technologies freely without fear of being sued for patent infringement. This would boost the amount of innovative activity in the area and effectively lower the cost of producing useful output for customers who use the resulting technologies. Intel has done this by creating "lablets" that work closely with universities to collaborate on research that will be published and not owned by Intel. IBM created a powerful example of this in its decision to transfer five hundred software patents to a nonprofit foundation in the open source community. IBM did something similar back in 2003 when it donated a bundle of development tools, called Eclipse, into the public domain. While IBM is being praised by the open source community for its generous donation of its intellectual property, IBM shareholders might also praise the company for helping lower the cost of the software on which IBM builds its own offerings. Instead of having to pay Microsoft or another company for a proprietary operating system, open source guarantees a cheaper alternative that will work well with IBM's products and services. IBM's generosity has the practical effect of lowering the costs of the operating systems that IBM sells along with its own offerings. This is a good way to boost your own business model.[30]

A related business model that makes good business sense is to be very open with technologies that are not inputs to the company but still are complementary to the core activities of the firm. Enabling many others to work with these technologies may expand the demand for the firm's core activities (which are not as open) and increase its profits. This is another rationale for Intel's corporate venture capital program, Intel Capital. Start-ups may develop uses for Intel technologies that were not known to the firm, providing the best market research money can buy. And increased demand for those complements will increase demand for the core technologies of Intel.

A third, more subtle, and perhaps even more powerful, strategy to leverage open source in one's business model is to develop system architectures that build on it. In a world with lots of useful building blocks, the ability to create and capture value shifts from developing yet another building block that is slightly differentiated from the others, to crafting coherent combinations of building blocks into systems that solve real commercial problems.

This competition took place recently in Web services.[31] Microsoft was trying to establish its .Net architecture as the platform for these services. That architecture sought to leverage Microsoft's tremendous franchise in its Windows operating system, and the extensive community of developers and other third parties that have based their livelihood on it. IBM, by contrast, countered with its WebSphere architecture, which had to work with Windows, but took the opportunity to leverage open source technologies—along with the community that has arisen around those technologies—far more extensively. And the key determinant in this competition ultimately depended on the decisions of the many independent software and services providers that made their own investments in choosing which architecture to support. These actors had to decide where to focus their own business model, and where the opportunities for value creation and value capture seemed to be the greatest.

Thus, while open source was created in ways that sought to deliberately eschew the creation of IP rights over its technology, alert companies have nonetheless developed business models that are propelling the technology forward into the market.

The existence of effective business models for open source augurs well for the further adoption of open source software. It is these companies' business models—not the rhetoric of "software should be free"—that compose the real threat to companies like Microsoft. And companies like IBM are developing business models that simultaneously exploit the availability of high-quality, low-cost open source software, even as they profit from greater patent protection for more proprietary technologies in other parts of their business.

While open source has been celebrated as a new and different approach to software development, its emergence ironically has coincided with the emergence of stronger intellectual property protection for patents and other IP. Alert companies will construct business models that incorporate *both* trends in their logic. In the following chapter, we will explore the stronger environment for patents and other IP and examine how this stronger environment influences the design of the business model.

3

The New Environment
for Business Models

In the previous chapter, we examined the challenges facing a firm as it tries to modify its innovation processes to become more open. We concluded that the firm's business model itself had to become more open for the firm to open up its innovation process effectively. In this chapter, we consider the world surrounding the firm developing its business model. As we shall see, this external environment is expanding rapidly, bringing new challenges and new opportunities to the creation of business models and the management of innovation. A new class of innovation markets is emerging, creating the beginnings of a secondary market for innovations and associated intellectual property. But to understand this potential future state of a world rich in markets for innovations and IP, it will be helpful to begin with a little history.

A BRIEF HISTORY OF PATENTS
IN THE UNITED STATES

On April 10, 1790, President George Washington signed into law the bill that provided the foundations of the U.S. patent system.[1] Back then, this patent system was unique; for the first time in history, the intrinsic right of an inventor to profit from his invention was recognized by law. Previously, privileges granted to an inventor were dependent on the prerogative of a monarch or on a special act of a legislature, all on a case-by-case basis. Protection for inventions was, in part, a manifestation of the U.S. Constitution, which explicitly provided that "Congress shall have the power . . . to promote the progress of science and useful arts by securing for limited times to authors and inventors the exclusive right to their respective writings and discoveries."[2]

In 1790, one Samuel Hopkins received the first U.S. patent, for an improvement in the making of potash (which is a substance derived from the ash of burned plants and is useful for making soap). Of greater interest is the fact that the reviewer of this first patent was Thomas Jefferson, the secretary of state, who was a rather accomplished inventor himself. Jefferson's review was not enough to cause Hopkins to receive a patent, however. Other signatures—from the secretary of war, the attorney general, and even President Washington—were required before Hopkins got his patent. Hopkins's patent application cost about $4.

By 1793, Jefferson stopped reviewing patents himself and assigned these tasks to a State Department clerk. The U.S. Patent Office itself was formed in 1802 during Jefferson's presidency. Initially, the patent system was simply a registration system that enabled inventors to have an official record of their patents. No attempt was made to see whether the claims of one patent conflicted with another patent, or whether the claims of the patent represented a material advance over the current state of the art. That changed in 1836, when the United States moved to a formal examination system to assess the novelty, usefulness, and nonobviousness of patent claims before granting a patent.

Throughout the nineteenth century in the United States, patents became more difficult to obtain, and more valuable when they were obtained. This was a conscious outcome of government policy to provide strong incentives for invention in the hopes of stimulating greater social advancement. Abraham Lincoln famously remarked that the patent system was intended "to add the fuel of interest to the fire of genius," and Lincoln himself received a patent.

While patent protection was considered to be very strong at the start of the twentieth century, the next eighty years caused the strength of patent protection to become progressively weaker. As U.S. corporations became larger and stronger, the U.S. government became increasingly suspicious of patents. Instead of fueling the fire of genius, the government worried that patents were being used as a means for monopolies to escape antitrust protections. Companies like AT&T and Xerox were sued by the government and forced to publish or license their patents to all interested parties.

Other patent holders found that patent protection often wasn't very effective as a means of excluding competitors from using their inventions. A common defense against a patent infringement suit was to claim that the patent was not in fact valid. From 1953 to 1977, only 30 percent of patents that were challenged in this way were subsequently upheld in court.[3] Patents were not regarded as particularly effective ways to protect inventions, and small companies were particularly disadvantaged, since the litigation process was very expensive for them.

In the 1980s, a backlash against this weakened protection emerged, in part due to the perceived Japanese threat to U.S. competitiveness in many industries. U.S. firms, some people thought, needed greater incentive to invest in innovation, which would be critical to restoring the U.S. competitive advantage in the world economy. This led to the creation of a new dedicated federal circuit court of appeals for patents. This court proved to be a more favorable court for patent owners, in that the court upheld the validity of patents in 68 percent of the cases during its first four years, a rate more than double that of the 1953–1977 period.[4]

This court even now continues to be much more "pro patent" than courts were in the earlier period. Indeed, the court has expanded the scope of what can be protected by patents and increased the value of remedies that patent holders can claim from an infringing party. This situation led to some dramatic court cases, as will be discussed later. More typically, it influenced the negotiations between patent owners and those alleged to infringe on the patents. Most of the time, these parties settle out of court. But the terms of these settlements are strongly affected by the small portion of cases that do go to trial. As patent holders gained more protection and larger awards, the terms of negotiated settlements before trial undoubtedly shifted toward the patent holders. These historic shifts have opened the door to much more active, even aggressive, management of IP.

COMPANIES REWRITING THE RULES OF MANAGING IP

This new approach to management of IP can be best seen by reviewing the experiences of companies that tested the strength of their patents in court cases. Of particular interest are the experiences of Texas Instruments, Polaroid, and IBM.

Texas Instruments' New Practice: Profiting from IP Licensing

Texas Instruments (TI) was one of the first companies to benefit significantly from the new state of affairs in patent protection. Jack Kilby was an early inventor of the semiconductor who worked at TI and assigned all rights to his invention to TI. Robert Noyce of Intel also received important foundational patents for his work in semiconductors, but Noyce's patents issued quickly, while Kilby's patent did not. Kilby's patent application was filed with the U.S. Patent and Trademark Office in 1959 and with the Japanese patent office later on. For twenty-nine years, TI fought with Japanese patent officials to get the patent to issue and finally succeeded in 1989.[5]

Intel created a wonderful business in microprocessors, and during the course of its business, it cross-licensed many of its patents with other large companies. It chose to compete in manufacturing and product design and received little direct monetary compensation for its IP. However, the access to other IP cleared the way for Intel's success, so indirectly it benefited a great deal from its IP.

By the time the Kilby patent issued in Japan in 1989, TI found that it had more direct ways to profit from its IP windfall. By now, the semiconductor business was a global industry. And this newly issued patent gave TI the right to exclude others from many aspects of semiconductor design if they didn't pay royalties to TI. While TI had cross-licensed many companies' IP in exchange for its own (including Intel, among others), there were many other companies, particularly in Japan and Korea, that had not signed any cross-licensing deal with TI. Armed with the Kilby patent and another key patent for planarization technology, TI proceeded to file suit with many semiconductor companies. Over the next several years, TI generated hundreds of millions of dollars from these patents. In some years, TI received as much as 50 percent of its entire corporate net income from royalties, most of which were due to these patents.[6]

For a semiconductor company to make as much money from its IP as from its designs and its production facilities was unheard of. This meant that TI was not only a maker of semiconductors; it was also an owner of valuable IP that was making a separate contribution to the company's profits. It would prove to be a harbinger of many subsequent semiconductor companies (such as ARM, Qualcomm, and Rambus) that made most or all of their money from IP, not from products.

Polaroid's Success: A New Level of Patent Protection

Another event that reflected the new strength of patent protection was Polaroid's successful suit against Kodak in 1989. Polaroid, with the breakthrough inventions of its founder, Edward Land, had pioneered a number of important technologies in instant photography. Land was not only a prolific inventor, he was also an astute businessman, and he filed for hundreds of patents for the inventions that he and his colleagues created in Polaroid's laboratories. Kodak was the leading company in film photography in the world and was looking to grow. An obvious new market for the firm was instant photography. Kodak's scientists had created an alternative technology for instant photography, one that they thought would not infringe on Polaroid's patents. It is likely that senior management within Kodak calculated that, even if the Kodak technologies did infringe to some extent, Kodak could negotiate a settlement with Polaroid that would enable Kodak to share the instant photography market with Polaroid. After all, patents had not been particularly valuable up until that point in time.

If indeed Kodak executives made this calculation, it proved to be an expensive error in judgment under the new, stronger patent regime. Polaroid filed suit for patent infringement. While the trial took place over many years, from the initial trial to the subsequent appeals, Polaroid won every important issue in the end. Polaroid alleged that the infringement was willful and that it deserved treble damages for the deliberate acts of infringement. To Kodak's disappointment, this was granted. Polaroid got the largest settlement ever awarded by a U.S. court to that point in time (over $900 million) for Kodak's infringement of Polaroid's many patents in instant photography. This stupendous sum far exceeded even Kodak's worst nightmare and reflected the new strength of patents. To make the matter worse, the court awarded Polaroid an injunction that forced Kodak to drop out of the instant photography market, restoring Polaroid's effective monopoly in instant photography. This raised the total cost to Kodak to more than double the damage award itself, because the company had to repurchase all of its unsold products from its distribution channels and write off all of its manufacturing investments in instant photography as well.

IBM'S EVOLUTION OF IP MANAGEMENT: FROM FREEDOM TO MONETIZATION

Other alert managers began to take note of the newfound strength of patents and altered their own IP strategies accordingly. A particularly impressive response came from IBM. Up through the late 1980s, IBM conducted substantial R&D and filed a number of patents in the course of its business. However, these patents were treated as something of an afterthought or as bargaining chips for cross-licensing with other companies in the computer and related businesses. As Ralph Gomory, who was then head of IBM's research division put it, "What we wanted was the freedom to invent."[7]

As evidence of the stronger value of patents emerged, IBM managers began to shift toward a much more aggressive patenting strategy, followed by a more proactive assertion policy. IBM rapidly increased its patent applications and rose to become the leading corporation in the world in terms of the number of U.S. patents it received.[8] IBM also began managing these patents as sources of financial revenue. Its royalty payments started at some few millions of dollars in the early 1990s but grew to $1.2 billion by 2004.[9] "We began to focus on how to monetize our intellectual property," says Joel Cawley, of IBM's corporate strategy unit.[10]

We will examine IBM's evolution in greater detail in chapter 8. For now, note how IBM has adjusted its IP management policies as the strength of patent protection increased.

THE EMERGENCE OF SECONDARY INNOVATION MARKETS

The experiences of TI, Polaroid, and IBM illustrate the emergence of an important force that is affecting the external innovation environment: the growth of what Ashish Arora and his colleagues called "intermediate markets," or markets for innovations.[11] Their term *intermediate market* refers to a market that emerges after the creation of a new technology, before that technology has been sold. In this intermediate market, ideas and technologies are developed by sellers and sold to buyers who take those ideas and technologies and sell them to consumers. Two quick examples of this intermediate market are mortgages and biotechnology drug development. In each instance, there is a group that originates the mortgage or the drug, and a second group that buys the mortgage or drug. In the latter case, the buyer invests additional money to take the drug to the market.

In the Closed Innovation model, companies had to take their new discoveries to market themselves for two reasons: first, because they would obtain more money that way; and second, because there weren't many other companies that knew enough to use the technology successfully. Intermediate markets for innovation in the Closed Innovation system were sparse. In an Open Innovation world, where useful knowledge is widespread, there are many companies with many potential ways of using a new technology, and many potential technologies that might be utilized in a company's business model. No company can hope to exploit all of the many ways a new technology might be used, so Open Innovation companies typically license technologies liberally to other companies. This creates a secondary market for innovations.

The presence of these secondary markets expands the number of ways a new technology can be used and promotes specialization among the different participants in the market, a division of innovation labor, if you will. Some companies specialize in creating new technologies, others specialize in developing new products, and still others focus on special niches, services, or applications along the way.

Secondary Markets in Products

As Arora and his colleagues found, a pronounced division of innovation labor has emerged in the chemicals industry. When new chemical plants are built, the company building the plant typically hires a specialized engineering firm to design the new facility. These specialized firms work on virtually all of the new chemical plants being constructed around the globe, so they are up-to-date on the latest ideas and techniques for making the plants as efficient as possible. Since these plants are extremely expensive, amounting to billions of dollars each, no individual chemical company builds them very often. So the specialized firms are able to accumulate knowledge and learning much faster than even the biggest of the chemical companies.

Another example of this specialization of innovative labor can be seen in the semiconductor industry. Back in the 1960s, the major semiconductor firms were captive subsidiaries of product firms, such as IBM or AT&T. There were markets for the final product systems, but no markets for the components of these systems. By the late 1970s, independent firms like Intel and Texas Instruments specialized in making chips and selling them to product companies, which used those chips to create new computer systems, or cell phones, or videogame players. Markets had emerged for chips, which were purchased and integrated to make systems products. By the 1980s, the manufacturing function in developing chips separated from the design function, as semiconductor fabrication companies (known as "fabs") like Taiwan Semiconductor Manufacturing Company (TSMC) built chips that were designed by so-called fabless companies, which effectively outsourced their manufacturing. Now there were markets for semiconductor manufacturing capability and associated markets for assembly, packaging, and testing capabilities. In the 1990s, companies like Qualcomm and ARM began selling intellectual property such as tools and designs to the companies that were designing and building chips. So now a company could buy a design from ARM, have it built by TSMC, and then offer it for sale, creating a specialized market for semiconductor designs.

Surrounding this vertical separation of functions in the semiconductor value chain are still more companies offering design tools, test equipment, and other services to the industry. This specialization has migrated around the world. In China alone now, there are more than six hundred specialized semiconductor design houses, and a number of new manufacturing facilities are being built as fabs for other companies around the world to build their chip designs.[12]

Yet another example of this innovation specialization comes from the life sciences. Thirty years ago, drugs were discovered, developed, tested, and marketed by large pharmaceutical manufacturers. By the 1980s, however, specialized biotechnology firms began to discover and patent new compounds. They would then partner with a pharmaceutical company that would take the compound through the clinical trials required by the Food and Drug Administration and then sell the drug to prescribing physicians. More recently, there has emerged a group of contract research organizations that partner with the biotech and pharmaceutical companies to conduct the clinical trials for them. In the 1990s, Millennium Pharmaceuticals began doing contract research for pharmaceutical clients, but reserved residual fields of use for a compound for itself and started to develop new drug applications for these compounds in 2000. Still other firms offer specialized equipment, tools, tests, and other services that assist in the drug development process.

This specialization of innovation also is emerging in the consumer products sector. Procter & Gamble has had a long tradition of great internal science, which it has used to create differentiated products to offer to its customers. More recently, though, P&G has realized that its core strengths are not in science, but in its ability to create strong brands. In some of its new brands, such as SpinBrush and Swiffer, P&G has created new and large businesses with technologies that it acquired from outside the company. Through its new innovation processes, which it calls Connect and Develop, P&G seeks to exploit the market for external technologies as it seeks opportunities to create new brands for its customers.

Secondary Markets in Services

This innovative specialization needn't be based on products per se. There are intermediate markets that have developed for services too. If you think back just a generation ago, you'll realize that a lot has happened to the mortgage industry in the United States.[13] From the early twentieth century through the 1970s, most people obtained a mortgage from their local bank. This was necessary because only a local bank knew the local market well enough to be able to assess the proper value for the house (and hence, the amount that safely could be lent against the property). The local bank could also assess the risks of the individual lender and service the mortgage appropriately. At that time, there were few "standards." Every mortgage was different, and every borrower had unique risk characteristics. So mortgage lending was a local business. Most mortgages were originated by the local bank, serviced by the local bank, and held by that bank until the mortgage was paid off.

This all began to change in the 1980s. Salomon Brothers, an investment bank in New York, realized that, under certain circumstances, one could create intermediate markets where mortgages could be traded after they were originated. However, certain characteristics of mortgages would have to be carefully defined, in order to know the risks of trading them. By defining an information bundle—which they took from the regulations established by the Federal Housing Administration (FHA)—the prior information asymmetries that confounded the ability of out-of-area lenders to fully evaluate the risks of locally originating loans were reduced to an acceptable level. Another important insight by Salomon was that individual mortgages' risk characteristics became more tractable as they were bundled together. These bundles could be resold as an investment asset to other parties.

This changed the basis of competition within the mortgage industry. Local knowledge became less important. The ability to access capital more cheaply now became more important in competing for loans. With more recent credit-scoring algorithms, that are based on all of a consumer's debt obligations and that provide more accurate measurements of her ability to repay her debt obligations, out-of-area lenders are at even less of an information disadvantage in lending. The mortgage industry's market structure has now changed dramatically. Seldom is a mortgage originated, serviced, and held until the end of its life by a single bank or other lender. Instead, there are specialist firms that originate the mortgages, other firms that purchase groups of these loans from the originators, still other firms that service the loans, and other brokers and advisers in the mix as well.

In countries like the United Kingdom, this evolution has gone still further. U.K. borrowers can link their consumer debt together with their mortgage, creating a secure credit facility with the holder of their debt. This reduces credit risks for lenders, because of the security interest in the home, and lowers interest rates for borrowers, as a result of the greater security. Current regulations in the United States have prevented this latest step from occurring here, but it is likely only a matter of time before most U.S. consumers have access to similar credit facilities.

Another services industry that has evolved a remarkably complex ecosystem of innovation is the entertainment industry. Back in the days of the Hollywood studio system, actors, directors, and other support functions were all organized within vertically integrated studios. Stars were under long-term contracts to the studios, which manufactured the rising stars' image and (hopefully) popularity, chose those stars' next acting projects, and handled distribution of the movies to their own captive movie theaters.

Fast forward to today, and every vestige of the studio system has vanished. Actors, directors, screenwriters, special effects specialists, and many other functions are organized from project to project, and a thick market of reputations and agents facilitate the process of selecting the next team of people for the next project.

One small example of the opportunities that exist under this new system comes from a brief history of the stage show *Chicago*.

Chicago was adapted from a book that was one of almost six hundred books that had been bought up by an investor group. The books were largely out of print, and nothing was being done to market them. The investor group, which included Richard Kromka, had the concept that some of these out-of-print titles might be redeveloped into valuable properties—a sort of IP real estate developer for the entertainment business. One key asset Kromka was able to attract was the participation of the actor Michael Douglas. Douglas's industry knowledge was important, but it was his ability to get a meeting with anyone in the entertainment industry that proved truly valuable.

Chicago was the first of these projects. The book was turned into a screenplay. The screenplay was shopped on Broadway to entice producer Harvey Weinstein. When the show got going, its obvious popularity caused Weinstein (with Kromka's group as a minority investor) to pursue a movie project as well. Douglas's contacts again helped here, and his wife, Catherine Zeta-Jones, agreed to play one of the lead roles in the production. Fueled by their portion of the success of *Chicago,* Kromka and his fellow investors are raising another round of financing to carry this model forward into other projects.

Another firm that is leveraging the knowledge and celebrity of one of its partners is Elevation Partners, a new venture capital investment firm. Bono, the lead singer of U2, has joined with Roger McNamee and others to form the partnership, which seeks to invest in new opportunities in the entertainment industry. Like Kromka's venture, the bet of Elevation Partners is not only about superior knowledge but also about superior access to the market: who in the entertainment world would turn down a chance to meet with Bono? This superior access enables Elevation Partners to pursue value creation from assets that other parties might not be able to commercialize as effectively.

SOME PRELIMINARY EVIDENCE ABOUT SECONDARY INNOVATION MARKETS

Despite the examples just described, the overall secondary market for innovations is in a very rudimentary state. That makes it challenging to provide much convincing evidence of its existence and growth, beyond supplying some examples. But in research I have conducted with my colleague Alberto Diminin, we have found what we believe to be some initial evidence of a growing overall secondary market for innovations.[14] This comes from our analysis of patent reassignments.

When a patent is initially granted, its ownership is assigned to the inventors of the patent. In most cases, the inventors in turn immediately sign over ownership of the patent to the company or university that employed them when they made the discovery. Usually, the patent is simply held by the owners until it expires, twenty years after its initial submission date.

Occasionally, though, a patent will be reassigned—that is, sold to another party. When such a reassignment occurs, the new owner notifies the U.S. Patent and Trademark Office (USPTO). In the case of patents, an assignee and an assignor have to complete a special form (PTO 1595) and to register the reassignment with the USPTO. A patent can be reassigned many times during its life, though this is rare. We analyzed the data from the USPTO on reassignments, using a database provided by Dialog.[15]

As figure 3-1 shows, patent reassignments are on the rise, both in absolute and in relative terms. There were fewer than three thousand reassignments back in 1980, while in 2003 there were more than ninety thousand reassigned patents recorded by the USPTO. The reassignments in 1980 amounted to less that .1 percent of all patents issued within the past seventeen years (and hence, available for potential reassignment), while the 2003 reassignments amounted to more than 4 percent of all patents issued within the past seventeen years. We estimate that a typical patent issued in 2003 has a 25 percent chance of being reassigned at least once during its life, based on the current trends.

Why are these patents being reassigned? On form PTO 1595, the USPTO asks the parties to specify the reasons for the transfer of the patent. These reasons include such diverse responses as correcting errors, assigning ownership to affiliated companies, mergers and acquisitions (where the ownership of the patent is being changed to keep up with the new owner of the company), transfers between independent companies, and an intriguing category called "security interests."

FIGURE 3-1

Patents being reassigned and published each year

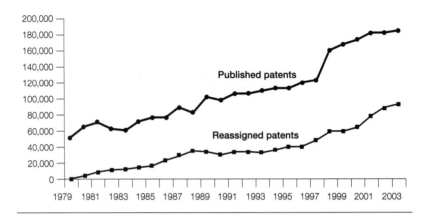

Looking at the reassignments in one industry, semiconductors, we collected all the reasons given on this form for why the reassignment had occurred. Figure 3-2 shows the reasons given in 1994. Twenty percent of the reassignments were between affiliated companies, another 14 percent were stand-alone IP transactions, and 15 percent were due to the use of IP as collateral for a loan, or "securitization." Figure 3-3 shows how these reasons had changed by 2003, as the industry consolidated. Affiliated transactions now account for more than half of all reassignments, at 61 percent, while security interests have risen to 23 percent.

Transactions between affiliated companies do not provide evidence of a growing secondary market because these are transactions between different parts of the same overall company (perhaps reassigned for tax reasons or for preparing for a spin-off of a subsidiary). But the growth in security interests in IP does seem to provide evidence of a secondary market. Here, IP is used to secure better loan terms from banks and other financial institutions. Typically, the IP remains with the firm, provided it pays back the loan. If it doesn't pay back the loan, however, the lender now has a claim on the IP assets and will want to find a way to convert that asset into cash. This will require a secondary market transaction.

FIGURE 3-2

Reassignments in semiconductors, 1994

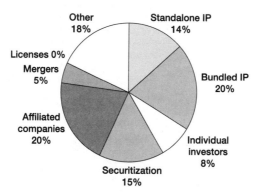

Total number of reassignments was 516.

FIGURE 3-3

Reassignments in semiconductors, 2003

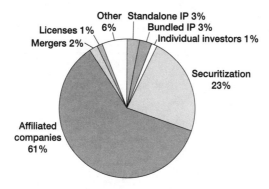

Total number of reassignments was 3,891.

Individual firms vary in their use of reassignments. Table 3-1 shows how the top nineteen firms in the IT sector (based on the number of reassigned patents that they were involved in from 1980 to 2003) varied in their use of reassignments. IBM had the most reassigned patents, but since IBM is also the most prolific patentee among

TABLE 3-1

Reassignments in the IT hardware industry, 1980–2003

Company name	Total patents reassigned to	Total patents reassigned *to* company (%)	Total patents reassigned *from* company (%)
IBM	40,443	2	5
Hitachi	33,372	7	1
NEC Corporation	21,756	2	2
Fujitsu	19,964	3	3
Hewlett-Packard	18,802	3	2
Motorola	18,654	3	28
Xerox	16,265	4	53
Lucent	13,938	3	27
Texas Instruments	12,599	3	4
Micron Technology	12,580	9	1
Intel	10,488	10	1
Corning	8,813	24	5
AMD	8,111	5	7
Alcatel	6,475	28	6
Ericsson	6,473	9	4
STMicroelectronics	5,592	13	0
Nortel	5,481	49	6
Sun Microsystems	4,679	6	0
Nokia	4,147	9	2

these firms, those reassignments are but a small fraction of IBM's total patent portfolio. Firms like Corning, Alcatel, and Nortel, for example, were active buyers of reassigned patents. Firms like Xerox, Motorola, and Lucent were active sellers.

This is characteristic of an active secondary market: there are some buyers, some sellers, some who are active on both sides of the market, and some who sit on the sidelines. Figure 3-4 provides one way to organize the data from table 3-1. It shows the balance of net reassignments (the difference between the ones each company bought versus the ones each company sold) on the y-axis. On the x-axis, it shows the level of reassignment activity in relation to its

FIGURE 3-4

Balance and intensity of reassigned patent trade in the IT hardware industry, 1976–2003

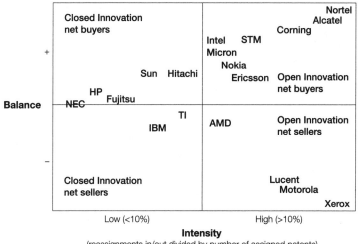

Low (<10%) High (>10%)

Intensity
(reassignments in/out divided by number of assigned patents)

overall patent holdings. While firms like IBM, Hitachi, and AMD were in the middle, Corning, Alcatel, and Nortel were building up their IP portfolios in IT through buying patents. Xerox, Motorola, and Lucent were liquidating their patent portfolios through selling their patents.

This data does suggest that there are at least the beginnings of a secondary market for IP. If this is indeed happening, how does this change the management of innovation?

MANAGING INNOVATION IN THE PRESENCE OF INTERMEDIATE MARKETS

While the intense specialization from intermediate markets has unleashed a lot of innovation in industries like semiconductors, drugs, and consumer products, it can be challenging to construct and manage a business model in the presence of these intermediate markets. When a company brings in an external technology to its business, for example, it must carefully assess whether it has the legal ability to use that technology without infringing on the legal rights of another company. This means that anyone scouting for external ideas and technologies must be alert to their legal status. As we shall explore in chapter 4, the protection of a patent for a particular technology is unlikely to cover every aspect of its usefulness. Even if the entity licensing or selling the technology has patented the technology, for example, the scope of that patent may or may not cover the uses that the acquiring firm wishes to practice. In turn, the protection of a technology may involve claims that inadvertently infringe on some aspect of another company's technology. (See figure 3-5.)

The fact that the required patent protection may not overlap entirely with the technology practiced by the innovating firm creates different situations for the innovator. Moving in the figure from left to right, the first region is one in which the patent provides protection over an area in which the technology is not practiced by the patent holder but may be practiced by someone else. The second region is "impaired protection," since the protection offered by the patent, in an area in which the technology is practiced, overlaps with the claims of

FIGURE 3-5

Patent scope, rival patents, and technology usefulness

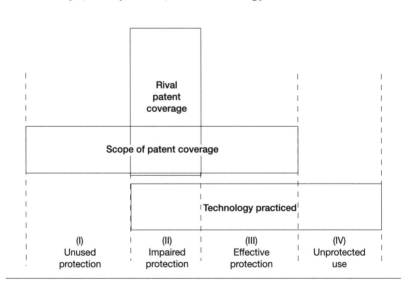

the patent held by another firm. The third region is protected use, where the patent claims do indeed offer protection for the practice of the technology. In the fourth region, "unprotected use," the technology may be practiced, but that practice enjoys no protection from the firm's patent. This region is extremely hazardous, since the practice of the technology may infringe on another firm's patent.

Shopping for the IP, as Well as the Technology

In a world of Closed Innovation, the analysis just provided would dictate where firms could enter, and where they should stay out of, the market in order to minimize their legal risks. In an Open Innovation world with developed secondary markets, the same analysis has a different set of implications. Now this analysis dictates where the firm should shop for rights to intellectual property that would fill in gaps in its own IP and that would support the use of external technologies within the firm's business. And in region I of figure 3-5, the firm might not only trade those rights to gain access to rights in regions II and IV; the firm might also seek direct compensation from firms that want to practice that technology.

So intermediate markets for technology in a world of Open Innovation profoundly change the management of IP. On the one hand, a firm cannot acquire and use an external technology unless it is confident that it has the legal right to practice the technology that it wishes to use (this would correspond to region III in figure 3-5). This ability to practice a technology also was a concern with technologies in a Closed Innovation world. But in that world, the company knew the entire history of the internal technology. In this more specialized world, where technologies flow across the boundary of the firm (perhaps multiple times), obtaining the ability to practice a technology without incurring an infringement action by another firm is more challenging because the full history of the technology's development is not as well known.

On the other hand, secondary markets provide the opportunity to greatly increase the utilization of internally owned technologies by offering them to other firms for use in those other firms' business. Not only does this increase the utilization of a given technology, but it also increases the scope, or the number of areas, in which a technology might be used. This can increase the total value of a technology, creating stronger incentives for new technology creation. It can also make innovation more productive, as each innovation finds more uses.

The Risk of Contamination

Secondary markets for innovations present other challenges as well. Before a company identifies a promising technology, it must interact with many companies and explore a variety of possible technologies to have any hope of finding a useful technology.

This raises an old, but very important, problem first noted by economist Ken Arrow, which is known as the Arrow Information Paradox: "I as a customer need to know what your technology can do before I am willing to buy it. But once you as a seller have told me what the technology is, and what it can do, you have effectively transferred the technology to me without any compensation!" And that is not all. If the customer discusses possible technologies with a would-be supplier, but decides *not* to license the technology and instead goes off and designs a technology internally, the customer may have contaminated itself with the knowledge of that supplier. Any subsequent internal development in a related area by the customer may be challenged by the supplier, which might allege that the customer stole the idea from the supplier without paying anything for it. If the customer is a very large company and the supplier is a very small company, this David-and-Goliath situation may make a jury very sympathetic to the small company, even if the large company developed its approach in a completely independent manner. In this instance, a jury finding of willful infringement by the defendant may result in treble damages, as occurred in the Polaroid-Kodak case noted earlier.

These issues pose daunting challenges for companies wishing to access external technologies. They can't simply invite people with those technologies to come talk with them, for the reasons just noted. Yet, in a world of Open Innovation, there is too much good stuff available on the outside to simply ignore. So then, how can companies identify potentially valuable external ideas?

SCOUTING FOR EXTERNAL TECHNOLOGIES

Every company that has embraced the logic of Open Innovation has had to develop answers to these issues. While the specifics of each answer may vary from one company to another, the principles of how companies scout for external technologies apply more generally.

One principle is that there is an enormous amount of public information now available about technologies from around the globe. Thanks to innovations in search technologies—from general search engines such as Google, Yahoo!, or Ask.com, to elaborate tools such as IBM's WebFountain—companies have a wealth of tools to identify potential technologies. As John Wolpert likes to observe, there is a wealth of useful technology that is "hiding in the open," provided you are thoughtful about how to look for it and recognize it when you see it.[16]

What Wolpert means is that while these tools are available to anyone with access to the Internet, they also require a thoughtful, creative mind to exploit them to their fullest. It may not be sufficient to rely on traditional processes for procuring external technology. For example, it may not be obvious what the right search parameters are to identify a promising technology to solve a particular problem. This was illustrated by the experience of Adrienne Crowther, then a vice president of Analysis Group, an innovation consulting firm. She said:

> We were working with a client who wanted to utilize Open Innovation ideas to find more potential technologies to feed into its business in the HVAC [heating, ventilation, and air-conditioning] business. They had some opportunities that they wanted to address and initially gave the job of identifying potential external technologies to their purchasing department.
>
> Well, the purchasing department apparently approached this company's usual suppliers and reported that none of them had anything useful to offer to the company. When we heard this, we decided to set up our own search process, starting with searching on the Web. While it took a few tries to find the right search terms, once we did that, we were able to identify a wide range of options for the client. We had

to examine hundreds of possibilities and filter them down to a manageable number. But we found some wonderful stuff among all the Web links. One turned out to come from a university consortium at a Big Ten university that the client did not know about. Another potential technology came from a professor working in the U.K. who also was not known to the client.

These two opportunities would likely never have been found by the purchasing department, but we were able to locate them in just a couple of days. We have followed up with both opportunities, and it looks like each of them is going to lead to a new business opportunity for the client.[17]

The Value Chain

The work of Eric von Hippel also reminds us that customers can be very important sources of innovation ideas.[18] The customer knows her own problems very well and often has to make significant adaptations to the technologies that she purchases in order to make them solve those problems. There is a great deal of information from these adaptations that point the way for further innovation in products and services. Sometimes, the additional technologies can be sourced directly from the customer. Other times, the technology may come from another company working with the customer. In either case, the fact that the solution is already being used effectively within the customer's organization suggests that there may be important value in the solution for other customers with similar needs.

Turning the idea around, companies are the customers of their own suppliers. So companies ought to give suppliers their own list of ideas to add to the suppliers' offerings. In other cases, the company may want its suppliers to take on additional functions that the company initially developed internally but would like now to see implemented within its supply chain. This is one of the consequences of companies going on "an asset diet" to reduce their fixed assets while maintaining their business revenues. By shifting functions to their supply chain, companies also transfer or offload the associated internal assets that formerly supported that function. The supply chain may be able to spread those assets across many customers, spreading the costs of providing the function more widely than the customer could do internally. For example, when IBM takes over a data center from one of its customers, it transfers the IT assets to IBM's balance sheet. And it can combine the IT needs of the customer for data storage with the needs of other customers for data storage. This combination saves money for both the customer and IBM.[19]

Managing IP in the Value Chain

Whether working with customers or suppliers in the value chain, though, the issue of IP remains an important source of friction in the exchange. Who will own the resulting product or service if we share ideas on how to innovate more effectively within a certain process? What rights will my supplier have to offer this solution to its other customers, some of whom may be my competitors? What rights do I have to offer this product or service to my other customers, some of whom may compete with this customer? And where do my rights end, and where do my customer's or supplier's rights begin?

IBM has had to sort out these issues in its implementation of its First-of-a-Kind program. In this program, IBM sends some of its own research staff onto the premises of one of its leading customers, which is tackling a challenging problem. Because the solution to the problem is not yet known (by design, it is a challenging problem), IBM and the customer cannot tell in advance exactly how ownership of the output will be legally partitioned between them. And since they are collaborating on solving the problem, it isn't entirely clear which should own what when an answer is obtained. The first time IBM engages in this process with a customer, these legal hurdles are for-midable. With time, however, IBM has learned some heuristics that have simplified the partitioning of the IP. At the simplest level, which-ever first thought of it, gets the rights to it. At a deeper level, though, whichever firm has the most compelling business model to exploit it, gets the rights to it (or, alternatively, there is some trading of IP rights that culminates in that outcome). Other outputs will be cre-ated jointly. These will be owned jointly so that each party can use the idea on a royalty-free basis. Occasionally, a joint committee will be created to oversee the subsequent use of jointly developed technologies.

Beyond the Value Chain: Business Networks

The business networks in which a company operates can also be a fruitful source of external possibilities. Informal sharing of information and knowledge trading can lead to the discovery of useful ideas that might solve important business problems. Larger communities where public information is exchanged, such as industry conferences and trade shows, also supply a great deal of public knowledge that can lead an alert innovator to useful solutions. These groups exchange substantial amounts of information, but this exchange is considered generally to be in the public domain. The most valuable information here is often where to look to obtain the location of private information, which then would have to be pursued under nondisclosure.

Technical standards bodies comprise another resource for accessing available knowledge about a particular technology and then forging a shared approach across a number of firms for how to apply that technology. Even here, though, IP issues surface with great frequency. These groups are not purely neutral forums, trying to develop the best technical solution to a particular technology problem. The research of Mark Lemley shows that technical standards bodies have a wide range of rules regarding how much IP must be disclosed to others in the technical body.[20] And this variation in rules can be leveraged by alert companies that position themselves to occupy key positions within an emerging standard.[21]

One such example is Rambus, a virtual semiconductor design firm offering a technology to speed up DRAM chips inside computers, which has profited significantly by exploiting loopholes in the rules of its standards-setting body. The body was created by companies that wished to increase the rate of data transfer between the memory and the system. After that body settled on a standard for how to accelerate the speed at which DRAM chips transferred data to the system, Rambus revealed that it had received patents on important elements of that standard. So any company that wished to implement the high-speed standard developed by the committee would likely infringe on Rambus's IP portfolio.

What Rambus did has been found to be entirely legal, in a series of court cases regarding its conduct and the legal rules around its intellectual property. The company's stock price is something of a pure play in that the intellectual property of the company is the only business it has. Therefore, Rambus's daily stock price reflects the market's current assessment of its value as a going concern, which is essentially the value of its business model and associated IP. The valuation of the company has experienced wide swings, from more than $100 to less than $10, even though the IP itself has been well publicized for many years now.

Leveraging University Relationships

Another vital source for accessing external ideas is to cultivate deep and ongoing relationships with universities. Faculty members typically are domain experts in fields that are potentially useful to many companies. Farsighted companies take the time to identify the academic thought leaders in the fields that they care about. Then these companies invest time to get to know these thought leaders, offering to visit their classrooms to help them teach and offering donated equipment, tools, and services to assist their research. Once a relationship has been formed, these thought leaders can help companies identify promising graduate students for summer internships and, later, possible employment offers. They can also help identify university research for later commercialization, as we shall see with UTEK in chapter 7.

University thought leaders can also play an important role for the company by serving on a technical advisory board (TAB). These boards are sometimes little more than window dressing for the company, to make its technology look more impressive. That is a waste of a potentially valuable resource. Some TABs play a more useful role for their companies by providing an independent perspective on technical trends and developments.[22] These companies review their long-term product development road maps with their TABs on at least an annual basis. External TAB members often know of other developments that the company itself had not heard of yet. Their views also may differ from the conventional wisdom inside the company about whether a technology is ready for commercial use. Unilever invites select university faculty into its labs in Manchester, England, on a periodic basis, where they wander around and talk to individual researchers about those researchers' projects. These highly informal contacts stimulate the academic, and the academic in turn occasionally connects one project within a Unilever lab to a different project that the individual researchers did not know about previously. On at least one occasion, that different project was in another Unilever lab!

Graduate students are another often-overlooked resource that can really help companies understand and use new research results from universities. By helping faculty sponsors execute their research, these grad students have learned a great deal about that work. They also work alongside many other graduate students, giving them a broad exposure to related ideas in the lab. The truth is, most graduate students are not well paid, nor are they treated very well at their universities. Many welcome the chance to work alongside respected industry colleagues to get a sense of what a future career might be like. Industry workplaces also feature the latest equipment and tools, which are not always available at the university.

FACTORS LIMITING THE EMERGENCE
OF SECONDARY MARKETS

As noted in chapter 1, the market for innovations already exists, but historically has been highly inefficient. While I have sketched the development of intermediate markets for innovation in a few industries, and provided some preliminary evidence that these markets are becoming more widespread, the fact remains that there are many inefficiencies that are limiting the emergence of secondary innovation markets. Understanding some of these inefficiencies allows us the ability to maintain a proper perspective on these markets. They also point the way to some mechanisms through which companies can overcome at least some of these current limitations.

One of the most critical limiting factors is the simple lack of information about the extent and terms of trade in secondary markets for innovations. Markets require information in order to function well, and much of the requisite information needed for coordinating market exchange of innovations is not yet available. For example, while there is an estimated trade of more than $100 billion annually in licensing for technologies, there is no place where this trade is reported and tracked.[23] What we know of the licensing market today comes from occasional surveys of companies (which ask the companies to report their trade in total) or from the occasional IP dispute in court, where the terms of a particular contract become part of the court record and made available to the public.[24]

The situation is somewhat analogous to the condition of the mortgage market in the United States before the advent of Salomon's bundling of mortgages. There are no information standards for technology licensing and associated IP trade. There is no FHA that defines a template or format for such trade. And given the wide range of terms and conditions for trading IP, it will continue to be difficult to aggregate statistics on this trade until one or more information standards arise.

Without this data, it is hard for companies to know what technology is available in the market. Business consultant Adrienne Crowther's experience for her HVAC client reveals both the potential and the problems of finding available technology. While she and her colleagues found two highly useful technologies in a short period of time, the client's purchasing organization was unable to find any useful technologies using its normal procedures for soliciting external inputs. This is typical of inefficient markets: you don't know what you don't know, so it is hard to tell what you may be missing. It will take experimentation with new processes to identify and exploit the opportunities latent in secondary markets for innovations. An emerging industry of "innovation intermediaries" has arisen to assist companies who wish to search more creatively and more extensively for external opportunities. We will explore some of them in chapter 6.

It is also very challenging to know how to value available technologies once they are located. Such value is determined by what a willing buyer would pay to a willing seller. Markets aggregate suppliers and customers, so any individual technology can go to the highest bidder and bidders know what similar technologies have sold for in the past, giving them a basis for calculating their bid price. But there is no systematic reporting of previous prices paid for external technologies and their associated IP. This makes it hard for sellers to know what price to expect to receive, or what price would be reasonable, given similar transactions in the past. So too for the buyers. Jeff Weedman of Procter & Gamble speaks of this as the "hopes and dreams" problem, where both sides to a transaction have unrealistic expectations, and there is little or no objective data to align those expectations more closely.[25]

POSITIONING YOURSELF TO EXPLOIT SECONDARY MARKETS

While it is too soon to say that secondary innovation markets have arrived in most industries in most advanced economies, it is not too soon to plan for the emergence of secondary markets in your industry. Secondary markets have had powerful impacts on industries—like semiconductors, biotechnology, consumer products, chemicals, and mortgage banking—when they do emerge. How can you assess whether and when secondary innovation markets are likely to impact your industry? Here are a few questions to explore to guide your assessment:

- Have any important technologies been introduced in your industry where one firm handed off the technology to another firm (via a license, joint venture, asset sale, or spin-off) at some point in the innovation process? How often did this occur last year?

- Did any university research projects turn into innovative new product or service offerings in your industry last year?

- How many patents were reassigned last year in the patent classes that are closest to your core technologies?

- How many times were you contacted last year with offers to license someone else's technology? How many times were you contacted about licensing your technology? How long did it take you to respond?

- Did any companies in your industry go bankrupt last year? What happened to their technologies and the IP associated with those technologies?

- How many sales and transfers of patents were you and/or your outside law firm involved with last year? How many sales and transfers was the law firm aware of in your industry?

- Are any new firms entering your industry with IP-based business models? (These are discussed in chapter 7.)

- Are any innovation intermediaries (see chapter 6) working for you or any of your major competitors?

- How many of your internal R&D projects were shelved or canceled last year without resulting in any licensing or spin-off activity?

Learning About the Secondary Market

If your exploration leads you to conclude that secondary markets have already arrived, or are about to do so, what steps can you take to plan accordingly? A first step is to invest in the development of market knowledge. Like any market, there are insiders and everyone else in markets for innovations. Take the time and invest the money to make yourself an insider. Elevate the importance of inbound and outbound technology licensing in your company so that it becomes visible to key business and technology executives. Compile a list of the transactions your company has already done on both sides of the licensing market. Encourage your top licensing executives to compare notes with peers in other organizations, and try to assemble some data on recent licensing transactions. Perhaps a university faculty member (and one or more graduate students) could be enticed to survey licensing activity in your industry. See what court records exist on terms and conditions of prior licensing disputes in your industry.

Making a Shopping List

In parallel with this market investigation work, develop a shopping list of ideas and technologies that would be of use to your company. Create a separate list of unutilized or underutilized technologies in your company that might be of potential use to other companies. This will allow you and your colleagues to engage with the secondary market as potential buyers and sellers. This will also give you more to talk about with people outside your organization, and you will learn more, faster. You might even find a deal or two in the process.

For many, another next step will be to identify one or more intermediary organizations that can represent you in searching for external ideas and technologies and/or help you shop your unutilized ideas to others. This is a great area to begin your exploration of secondary markets, and chapter 6 in this book can help you get started.

Preempting the Trolls

Another reason to engage with secondary markets for IP is to preempt the possibility that one or more "patent trolls" may come knocking on your door, demanding royalties for patent infringement.[26] Since trolls do not create their own technology and IP, they must obtain it from outside their organization, in the secondary market. If you actively engage in those same markets, you can acquire the IP (or a non-exclusive license to the IP) for a tiny fraction of what a troll would charge you later on to access that same IP. So send someone to those bankruptcy auctions of failed start-ups in your industry. Monitor the research activities of universities and research institutes. Cultivate relationships with individual inventors if they are working in areas that could relate to your own.

By fishing in the same waters that the trolls must fish in to obtain their IP, you can pay wholesale prices for IP, rather than retail. And once you are connected to the secondary market for IP, you may be surprised at the opportunities you discover to sell some of your unused IP to others (or perhaps to cross-license instead).

Remember the metaphor of the troll. He lurks under bridges or other choke points, preying on unsuspecting travelers. If you have identified the potential choke points in advance and have taken the appropriate precautions, the troll cannot catch you unawares. The next chapter will identify some patent-mapping exercises you can do to identify potential choke points, which can help you prepare for and perhaps preempt the troll.

THE NEW ENVIRONMENT FOR BUSINESS MODELS

Business models are challenging to develop, and effective business models are a tremendously valuable asset to the company. Intellectual property can play an important role in the development of an effective business model, and indeed many companies have enjoyed great success in recent years as a result of their IP management. Successful business models can create inertia as well, making it harder for very successful firms to respond appropriately to changes in their environment.

In this chapter, I have sketched out the new environment in which business models must be constructed and managed. Patents have become stronger, making possible the further specialization of innovation. New kinds of business models are being forged in this new environment, with companies that play by different rules than the established firms in their industry. We will examine some IP-enabled business models in chapter 7.

This new environment will pose new opportunities and new challenges, even to business models that are currently very successful. Companies will enjoy many new ways to enter into a particular business, by specializing in particular portions of a value chain. Other companies will find themselves vulnerable to some of the specialist firms and will see that their cross-licensing strategies that obtained design freedom in the past no longer work against specialist firms, such as trolls that do not wish to practice IP, only to license it. We will explore this at greater length in the next chapter.

4

The Impact of Stronger IP
on the Business Model

In the last chapter, we saw how patent protection had strengthened considerably in the past twenty-five years. This creates a
new environment for business models and suggests that every
company needs to pay closer attention to its IP than it used to. Protecting ideas is costly and time consuming, but it has become too
important to innovation to neglect. To complicate matters further,
there are new players emerging whose business models are focused
on extracting value for any of their IP that may be involved in your
innovations. And there may be instances where your business model
may dictate that you *not* protect all of your ideas, in order to create
value for your customers and collaborators.

In this chapter, we will explore how to link your IP protection to
your business model. We will also explore how you can leverage underutilized IP coverage that you have to either enter into new markets or
obtain revenues from others in those new markets. And we will consider how an Open Innovation approach to IP might preempt some
of the threats from firms specializing in IP licensing.

We will begin by looking at individual patents and technologies.
Then we will examine value chains and patent mapping, which connect a number of patents and a number of technologies. Then we

will consider the role of the technology life cycle in how to manage patents and IP. While many companies have grappled with these concerns, two things are missing from the ways that most companies manage IP. First, companies must connect the management of their IP to the underlying technology life cycle of that IP. Second, companies must change the management of intellectual property surrounding the technology in different stages of the technology life cycle. Those two often-overlooked aspects are explored here.

PROTECTING INDIVIDUAL PATENTS
AND TECHNOLOGIES

Choosing how to protect one's technology and ideas is a challenging and complex activity. There are many legal and economic considerations that must be taken into account.[1] Before developing the connection between technology, the business model, and IP management, it is good to develop some foundational concepts. I will begin with what patents do and do not protect.

While it is well known to legal experts, most managers do not realize that patents do not directly protect technologies. Patents may cover aspects of a technology that are embodied in a product. But the technologies in the product, and those used to make the product, may not align entirely with a company's patents. For the sake of what follows, I am assuming that the patents in the analysis are valid. (In many patent infringement cases, however, the defense usually claims that the patents in question were wrongly granted and therefore not valid.)

To illustrate what protection patents provide, consider the series of figures, figures 4-1 through 4-3.[2] In figure 4-1, I have created a schematic representation of a technology and its associated patent protection. The two are drawn deliberately such that the practice of the technology, and the protection for that technology, are not entirely aligned.

There are three regions of interest in the figure. The middle region is the region where the patent coverage and technology coverage overlap. Uses of the technology in this area are protected by the patents held by the firm. This is the conventional assumption made

FIGURE 4-1

Evaluating technology alignment with patent coverage

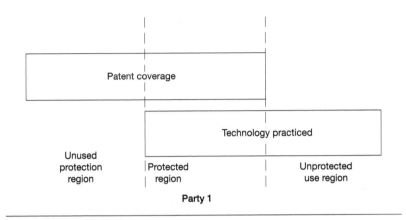

Party 1

by most managers—that they can practice their technology safely, due to their patent protection. (In practice, the protection may be provided by many patents, rather than just one. I suppress this to simplify the presentation of the argument.)

The region on the right, though, is a region where the technology is practiced (i.e., it is being applied to solve a real problem) without any protection from patents held by the firm. This is an unprotected region. Companies deploying a technology in this space are running a risk that someone somewhere else might have a patent that covers (or "reads on," in industry parlance) that usage. The left-most region is also typically neglected by most managers. This is an area where the patent provides coverage, but where the technology is not currently practiced. This can be thought of as "unused protection" because the scope of the patent coverage extends beyond what the company is currently using. In contrast to the unprotected region, this is a region of latent potential value, because that coverage could support extensions of the technology in that area or might provide licensing opportunities to other firms that might be operating in that space.

Figure 4-2 introduces a second party into the analysis. That second party also has some patent coverage and a region where its technology is useful. Like the first party, the alignment between the patent coverage and the usefulness of the technology is not complete. For ease of exposition, I have drawn the second party's position to be symmetrically opposite that of the first party. This makes the analysis much easier to explain, but each party's position is unlikely to be so symmetrical in reality.

The presence of the second party forces us to be clearer about what protection patents actually provide. A patent is a legal right to exclude others from practicing a technology when you own a patent that covers that technology. It does not actually grant you permission to practice the technology yourself. In order to have what industry calls "freedom of action," you must be sure that no other companies' patents cover your technology.[3]

Figure 4-2 illustrates this situation. The middle region that formerly was protected for party 1 is now transformed. The presence of party 2's patents now makes this region impaired for party 1. That is, party 1 cannot simply practice its technology as before in this region because party 2 also holds valid patents that read on this area. Party 2 may have the ability to block party 1 from using its own technology here or may choose instead to charge a fee for allowing party 1 to use its own technology.

FIGURE 4-2

Complex technology alignment when two parties have conflicting patent claims

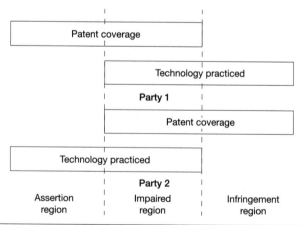

The right-hand region that formerly was unprotected has now changed to an infringing region for party 1. Party 1's practice in this region directly infringes on party 2's patent rights. Party 1 is vulnerable to party 2's enforcement of those rights.

The left-hand region that formerly was unused protection has now evolved into an assertion region for party 1. That is, party 1 has valid patent claims here that cover party 2's practice of party 2's technology. Here, party 1 has the choice of how it wishes to assert its rights and can charge a wide range of monetary amounts to enforce those rights (recall how TI much profited from its patents in chapter 3 by licensing them to other firms). However, the firm does not have to license. If it prefers, party 1 can simply deny party 2 the right to use party 2's technology in this area (as we saw in Polaroid's victory over Kodak, in which Polaroid forced Kodak to exit the instant photography market).

The stage is set in figure 4-2 for a grand bargain: party 1 and party 2 agree to cross-license each other so that both can practice their respective technologies without having to fight patent infringement litigation with each other. This cross-licensing is a common practice in complex industries like semiconductors, where every company must use the technologies of many other companies to make products.[4]

Now consider a slight modification to figure 4-2, shown in figure 4-3. Here, party 2 has the same patent claims as in figure 4-2, but in this case, party 2 is not making any products and is not practicing any technology. This subtle change dramatically alters the circumstances of figure 4-2. There can be no grand bargain here in figure 4-3 because party 2 is not at risk for losing the ability to practice its technology. Party 2 is a pure-play IP company whose business model focuses exclusively on IP and nothing else. Examples of such pure-play IP firms include ARM or Rambus in the semiconductor industry, NTP in litigation over Research In Motion's BlackBerry, or Dolby in the consumer electronics and entertainment industry. Other examples also include the so-called patent trolls that we discussed in the previous chapter.[5]

The pure-play IP business model does not value a cross-license; it wants cash compensation instead. In figure 4-3, this is shown in two ways. First, the infringement region for party 1 expands to include the

FIGURE 4-3

Complex technology alignment when second party holds IP, but does not practice technology

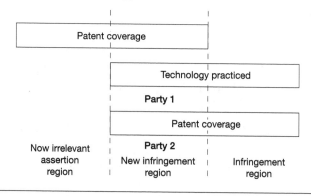

middle region. And second, the previous assertion region for party 1 is now irrelevant because party 2 is not practicing any technology covered by party 1's IP. This is a far more dangerous business model to confront for party 1 because its earlier ability to barter access to others' IP with its own IP is no longer effective. Correspondingly, this business model potentially increases the economic value of party 2's IP by removing any commercial activities that could be held hostage (assuming that there was some risk of infringement from those activities) in a negotiation with party 1.

MAPPING YOUR SITUATION

How does a manager know what his or her situation is, with regard to these different regions? The basic tool used to assess these questions is called "patent mapping." This mapping examines all of the granted claims of an issued patent and considers where they might apply. It then examines other patents from other patent holders to see where their claims might apply. This is an expensive exercise, and the resulting analysis is not nearly as neat and precise as figures 4-4 and 4-5. It has become increasingly essential, though, in creating and managing business models in many different industries.

FIGURE 4-4

A patent map of the value chain

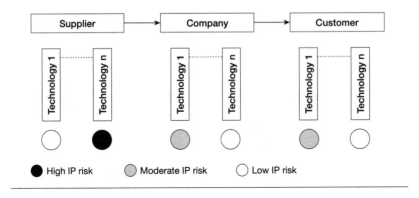

The earlier analysis looked at one technology in isolation. A more realistic analysis, and one that is increasingly common in many companies, is to map the entire value chain for a company.[6] At a minimum, such maps examine the company and its direct competitors, and then include the company's suppliers and customers (and often those customers' respective direct competitors as well). Figure 4-4 provides a simplified illustration of such a patent map.

In essence, figure 4-4 shows where companies are protected or exposed in different parts of the value chain. The earlier analysis of potential patent infringement of a technology being practiced in figures 4-1, 4-2, and 4-3 would be performed at each node of the value chain in figure 4-4. For simplicity and ease of communication, companies will often use black, gray, and white (or, equivalently, red, yellow, and green) to indicate the level of exposure that they face at each stage of the value chain. Black areas in the value chain indicate areas where the continued practice of the technology is at risk of infringing one or more patent claims of another company. Gray areas indicate areas of caution, where there may be some risks or challenges. White areas indicate areas of freedom, where the company can continue to operate with little or no risk to its business. In figure 4-4, the company itself faces a moderate risk on one technology (technology 1) and a low risk on another technology (technology n). And its customers

FIGURE 4-5

IP mapping value chain analysis: printers

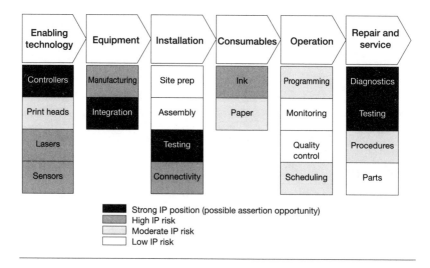

Strong IP position (possible assertion opportunity)
High IP risk
Moderate IP risk
Low IP risk

and distribution channels face a similar level of risk. However, there is a serious IP risk in the company's supply chain, as shown by the black circle for technology n. If the supplier is challenged on this technology, the supplier may not be able to continue its supply of that item.

Figure 4-5 shows an actual patent map of a value chain of printers, done for a company that makes specialty materials that are used in the printing market. It shows that this organization faces a number of areas where it is potentially exposed in different parts of its value chain. Some of the risk lies within its own business. But other risks lie within its supplier base (such as sensors and lasers, where those suppliers could be held up or even shut down for alleged infringement) or within the firm's customers (especially in the consumables area) and distribution channels.

These maps can reveal opportunities, as well as risks. There are areas in figure 4-5 where the company enjoys a strong position, primarily in integration and testing, which might be opportunities for product line extensions or external technology licensing. These could be leveraged to enter into adjacent markets, to generate revenue from licensing, or to neutralize some of the risk areas elsewhere in the value chain. And these are not mutually exclusive possibilities; the company could also combine two or more of them. The firm could also secure better terms from suppliers if it could provide some IP coverage to those suppliers in areas where the suppliers are exposed. Similarly, IP rights could become additional items in negotiations with customers and distribution partners to achieve better terms.

The clear message of figure 4-5 is that firms should not only think about their IP situation in their own business but also consider IP risks in other portions of the value chain.

Patent maps are snapshots of a company's situation at a single point in time. These proactive approaches become even more interesting in a dynamic context, which will be explored further in the next section. The question of whether to protect ideas, and how much to protect them, will be shown to vary over the technology life cycle.

THE DYNAMICS OF IP MANAGEMENT: LINKING IP TO THE TECHNOLOGY LIFE CYCLE

One thing that is well known and well accepted among those who study or work in technology is that technology changes rapidly. Some have defined a technology life cycle (TLC) to help describe the underlying patterns in the constantly changing world of technology. What has not been done to date is to link the management of IP and the business model to the TLC.

The proper management of IP should vary with the stage of the relevant technology in the TLC underlying that IP. Currently, most companies use a one-size-fits-all approach to managing IP, effectively ignoring the underlying technology life cycles. This is a mistake. IP management should align with the stage of the technology in question and help shape the subsequent stages of that technology's development. First, we will examine the life cycle concept in some detail and introduce the IP life cycle model. Then we will explore the implications of this model for the differing ways to develop business models and manage the intellectual property associated with the technology.

The technology life cycle is one of the most fundamental aspects of managing technology. The idea goes back to the seminal research of William Abernathy of Harvard Business School and James Utterback of MIT, and has been widely embraced by other scholars since.[7] This line of research has shown that technology does not simply develop at a single, straightforward pace. Instead, there is an initial period where a wide variety of technologies vie for acceptance in the market, a later period where the winning or "dominant design" technology establishes itself in the market, a third period where the technology matures, and a final period in which it becomes obsolete.

The notion of technology life cycles can be well explained with a simple S curve (or "logistic curve," as it is sometimes called). In figure 4-6, the performance of a technology is graphed on the y-axis, while the x-axis shows the time since the technology first appeared. Four distinct stages in the technology's evolution may be observed.

In the first stage, the technology is just beginning to emerge. Its rate of improvement is modest at best, and it takes some amount of time to eke out even this modest performance advance. This stage comes be-

FIGURE 4-6

Stages in the technology life cycle

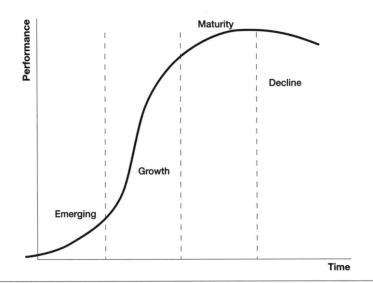

fore the emergence of a dominant design, so there are a wide variety of potential technologies. In automobiles, for example, there was a time in the very early 1900s when gasoline engines vied with steam and electrical engines to be the dominant power plant for motor vehicles.

The second phase is the growth phase; this is where technologies catch fire and grow explosively in performance. In the terms of academic scholars, the "dominant design" of the technology has taken root, causing the industry to focus its innovative efforts on how to advance the technology within the dominant design. Customers now find the technology to be a powerful solution for at least one pressing problem they face, and companies begin to incorporate the technology into portions of their business. The market then takes off, generating very rapid growth in units and in revenues. To stick with the automotive example, the gasoline engine became the dominant design for the power plant of the car, so innovations began to focus on how to increase the horsepower of the engine. Henry Ford's Model T, and the later Model A, are examples of products in this phase.

The third phase of the curve is the mature phase, where the market growth slows down and the industry reaches its maximum levels of revenue for the technology. At this stage, the technology is well understood, the leading players are well established, and there is little or no entry by new firms into the industry. The technology itself begins to subdivide, creating new niches of opportunity for novel applications of the technology, and penetration is now widespread around the globe. Within the auto industry, this might correspond to Alfred Sloan's segmentation of the auto market into different categories of vehicles and price points. More recently, the rise of sport-utility vehicle and minivan categories could be seen as outcomes of a mature phase of the industry. Customers now show real interest, as the technologies are proven and reliable. Customers themselves are able to innovate by creating new ways of applying the technology to their business problems.

The fourth phase is the declining phase of the technology. In this phase, the technology's rate of performance improvement reaches a plateau. Often, a newer technology has arisen to take the place of the earlier technology, substituting for the functions performed by the earlier technology. While cars continue to grow and advance, albeit slowly, there are examples of declining technologies within the industry. One such decline was clearly seen in the response of U.S. tire manufacturers to the advent of radial tires coming into the industry from Europe. Relative to the performance offered by radials, bias-ply tires simply couldn't advance much further and eventually conceded the entire market to the newer belt technology.[8]

This view of a technology life cycle is necessarily a simplification of the complexities of technological evolution. But the model shows some important features that will be important for the analysis of crafting business models and managing IP. One important feature is the necessary emergence of a dominant design to spur the growth of the technology in the market. Another important feature is that technological performance is of critical importance early on, while quality, volume, and cost become more critical to success later on. A third important feature is the high number of newly entering firms in the first two stages and the lack of such entries in the final phases.

Perhaps the overarching implication, though, is that companies cannot manage technology in the same way throughout each of the phases of the cycle. The key success factors for the technology in one phase are *not* the key factors in a different phase. Instead of using a one-size-fits-all model of technology management, a company instead must adapt its management of technology to the phase of the cycle in which it competes.

THE IP LIFE CYCLE MODEL

The very same logic of avoiding one-size-fits-all needs to be applied to business models and the management of intellectual property as well. Just as one should not manage technology in the same way in each phase of the technology life cycle, so too one should not manage IP covering that technology the same way throughout its legal life. As with the management of technology, the management of IP should be tailored to the phase of the technology life cycle that the IP covers. I call this tailored approach the "IP life cycle model." Here is how the model works:

- In the initial stages of a new technology, companies must invest in creating IP and choose the best method to protect that IP. That choice will depend on the role that the new technology might play in the company's business model.

- In the next phase, the company will deploy the technology and go to market. Options to take the technology to market include partnering to obtain distinctive capabilities required to complete the offering and accessing necessary complementary assets to support the offering.

- In the third phase, the company will consider how to harvest the fruits of the technology. While this certainly includes using the technology within the company's own business, the possibilities extend well beyond, to include competitors, customers, suppliers, and third parties in other markets. There are both revenue and profit opportunities that promote external licensing or spin-offs in many cases, as well as strategic considerations that might limit the outside licensing of the technology in other cases.

- In the final phase, the company will manage its exit from the technology. This might be forced by the expiration of legal protection for the IP (note that while patents and copyrights expire, trade secrets do not). Or the exit might also be motivated by the introduction of a new, improved technology that replaces the earlier technology. In these cases, the legal protection for the IP may extend well beyond its value in a use within the original business model. However, the IP may remain quite valuable to another company's business model in a different use.

Let's explore this process in more depth and see how IP might be managed uniquely within each stage.

Emerging Phase

In the initial stages of a new technology, the management of IP will depend greatly on whether the technology fits well with an existing business model. If it doesn't, the business model that makes the best use of the technology may not be apparent and will have to be discovered.

In the former instance, the management of IP is straightforward. The company should seek as much IP protection as it can afford, and it should aggressively develop the technology. When Texas Instruments developed its microprocessor patents, which we saw in the previous chapter, it already had a strong original-equipment-manufacturer (OEM) business model to sell chips to the consumer electronics and computer industries. TI could make these investments with the knowledge that the other elements of its business model already fit well with the technology. Important aspects of the business model—such as the distribution system, the manufacturing and operational processes, and the service and support elements— act as complementary assets in this phase. The brands and trademarks already owned by the firm may extend to cover the new technology. The fit with such complementary assets provides additional assurance that the company will be able to profit from the investment in the technology.[9]

When Apple began the development of its iPod player, it faced many risks. Could its player compete with the other MP3 format players that were already out there? Could its iTunes store, which used a proprietary format to download music, become the preferred place for content owners to list their titles? If Apple could engineer a better user experience, would its more proprietary business model (when compared to MP3 players using free music services that downloaded music illegally) fit with downloadable music?

Apple essentially bet that the answer to these challenges was yes. The company's business model is centered around offering a superior user experience as a value proposition. Apple knew from its many years in the PC business that its Macintosh technology was easier to use and easier to learn, and that Mac users believed they were receiving a superior experience. Apple could bring a similarly superior user experience to downloaded music, played on a handheld device. Moreover, its iPod used many other elements of its business model in operations, distribution, marketing, and sales.[10] Its IP is strongly protected, since Apple can keep the internal elements of its solution proprietary, enabling it to operate with trade secrets, patents, copyrights, and trademark protections. Apple's surrounding complementary assets in its business model provided further protection, which was one reassuring aspect of its business model to the music companies that would provide the content.[11]

But there will be other times when a technology does not fit with any established business model and a new business model must be created to commercialize the technology effectively. When Kodak first started working on digital photography, it quickly mastered the technology but struggled to figure out where it could make money. There was no equivalent to the film that Kodak used as its primary source of profits from chemically based photography.

Closer to the iPod space, RealNetworks developed some excellent technology to act as a streaming media player, able to play both MP3-formatted music and digital video content. Its RealAudio and RealPlayer, respectively, enable users to play music and video in a variety of formats. RealNetworks was a start-up and had to figure out a business model that would best allow it to compete with its technologies. The company experimented with a variety of approaches, from selling the standalone player, to bundling it in with subscriptions, to providing specific audio and video content, to creating radio over the Internet, and, later, to enabling computer gaming over the Internet.

In the beginning, Real needed to establish itself as the first and biggest streaming media technology and to try to become the industry standard. Its IP strategy was to brand its player and open up its interfaces while retaining proprietary control over the internal code. But the key strategic goal was to get its RealAudio software established as the de facto standard (i.e., the dominant design) for streaming audio content, and later, to get the RealPlayer software similarly ensconced as the de facto standard for streaming video content.

In this emerging phase, when a company lacks an appropriate business model at the outset, and there is no dominant design that determines how the technology will unfold, IP protection is secondary to the strategic mission of finding the business model that can best commercialize the technology. Indeed, a company often does not know the best use of a technology at the outset, and so it may be unclear how best to protect that technology for that use. There is little value in very strong protection of a technology that has lost the race to be the dominant design.

Growth Phase

In the next phase, the company will deploy the technology and go to market through a business model. If the company's technology wins the competition to be the dominant design, its technology and business model will likely be widely copied throughout the industry. Winning this competition requires successfully practicing the technology and developing the considerable know-how required to incorporate the technology into the customers' own business, often fitting the technology into a larger system. Significant tacit knowledge is built in doing this.

If the company's technology has not won, there is strong research that shows that it usually is futile to hang on to the approach that lost the race to be the dominant design.[12] Instead, it is best to either withdraw from the industry altogether or shift over to the winning design. In the latter case, the company needs to access the IP of the dominant design, as well as the associated know-how needed to practice the technology effectively. While reverse engineering and hiring away employees from the winning rival firm may help, the losing firm's IP may also be enlisted here. That IP may provide a bargaining chip to gain a license on attractive terms to the winning technology. Alternatively, the losing firm may prefer to exit the industry and receive some licensing revenues for its IP, a sort of consolation prize for losing the race to be the dominant design.

For the winning technology, intellectual property management in this phase takes on a dual character. The first portion of IP management was to support the company in its ability to make its technology become the dominant design. This may have caused the firm to consciously share much of the technology with other firms, in hopes of recruiting more firms to rally around its technology. In this second phase of the IP life cycle, the second portion of IP protection comes into play. Here, the firm seeks to capture a portion of the value created by the now dominant design.

There is a natural tension between these two elements of IP management in this phase.[13] As has been observed in many technology competitions, one can win the battle of becoming the dominant design and still lose the war to capture the value from this victory.

Rambus, the pure-play DRAM IP company we saw in the last chapter, started out by trying to set a standard with its technology. Only after that effort did not succeed did it shift to an aggressive licensing model to generate its profits.

The IBM personal computer is another example of this tension. The PC triumphed as the architecture that defined the evolution of personal computing since its introduction into the market in 1981. And for the next few years, IBM watched its architecture move its market share past Tandy, past Commodore, and even past Apple to become the dominant share supplier of PCs by 1985.

IBM also had crafted significant protection of key portions of its intellectual property for its PC. It had a strong brand, both for its corporate identity and for its PC product (those readers past a certain age may recall the Charlie Chaplin–esque, "little tramp" advertising theme). It also had patents on many elements of its computer design and carefully fenced off its ROM BIOS code that connected the DOS operating system with the PC hardware. As it happened, companies such as Compaq and Phoenix Software realized that they could legally reverse-engineer this code without violating IBM's intellectual property. While this was expensive and time consuming, once they succeeded in this effort, dozens of companies (which bought ROM BIOS chips from Compaq and Phoenix Software) could now offer truly compatible computers to IBM.

Thus IBM won the battle of becoming the dominant design. In 2004, however, the company threw in the towel and sold its PC business to Linova, a PC manufacturer based in China. The war *within* the IBM PC industry architecture over the long run had been won by others, such as Dell and HP (which bought Compaq). IBM's efforts to manage its IP succeeded in establishing the dominant design. Those efforts failed to enable IBM to protect enough of its technology to retain industry leadership two decades later.[14]

The iPod must also strike the right balance, now that the device is strongly in its growth phase. If it loses control of its architecture entirely, it could face the fate of the IBM PC. If it keeps too tight a grip on its proprietary approach, other, more open approaches will eventually overtake it in the market (as happened back in the 1980s with the Macintosh versus the PC).

Mature Phase

In the mature phase, the bulk of the market growth has already occurred. Instead of positioning itself for the future, the company must now focus on how to harvest the fruits of the technology today. As product lives shorten in specific markets, maturation encourages companies to begin to entertain other uses of the technology in newly emerging segments within the industry, and even in wholly different uses in other industries.

In the technology life cycle literature, this is the juncture where the basis of technical competition shifts from superior product technology to superior process technology. In part, this is due to the increasing difficulty in differentiating one's products from those of rivals. In part, however, there is now enough widely shared information throughout the industry that vertically integrated organizational structures confer fewer benefits, while more specialized and focused parties within the value chain (with highly optimized processes for those elements of the value chain) become increasingly competitive.

This phase ushers in a variety of innovations in business models, even as the technological differences diminish. Companies now craft many alternative business models, often involving new arrangements with competitors, customers, suppliers, and third parties in other markets. In the PC market, for example, the Dell Direct model emerged to become a dominant force as the industry matured. Taiwanese companies began to manufacture PCs and laptops to the specifications of the big-name computer manufacturers. Other companies started to reverse-engineer the printer cartridges of successful companies like HP.

This phase also ushers in a different objective for managing intellectual property. It is no longer enough for IP management to establish and defend the company's business model within the industry. IP management must now support the application of IP to new segments within the current industry and seek out new applications of the IP in other industries. Polaroid, for example, pushed its instant photography into many commercial applications, such as identity badges and security documents. Kodak for its part pushed its film technology into many parts of medicine, to provide high-quality images for patient diagnosis.

In many cases, those applications will not be pursued by the company owning the IP but instead will be realized by other companies. Overly restrictive IP management strategies will impair the search for these new uses of the company's technology. Proactive and appropriate strategies, however, can help the company identify and exploit these new opportunities. Business models for these new areas of application must take account of the value provided by the new application partner, as well as that of the technology provider.

In later stages of its evolution, RealNetworks has moved increasingly into content and services, now relying on exclusivity in access to content for particular display media (such as PCs and wireless devices such as cell phones). As streaming media has matured, Real has moved its player software (as opposed to its content) toward an open source approach. This increases the number of developers building on Real's technology, even as it loosens Real's ability to control the direction of the RealPlayer software technology.

It is likely that Apple's iPod will have the opportunity to be useful in a variety of applications well beyond the playback of music and video. But Apple will have to be quite agile and adaptable to find the right mix of protection and openness to penetrate these newer application opportunities in a profitable way.

Decline Phase

In the final phase of the technology life cycle, the technology no longer has much, if any, value in the original market. This might be forced by the expiration of legal protection for the IP. More commonly, though, in rapidly advancing industries, the market value of a technology expires long before its legal protection lapses. This typically occurs when there is the introduction of a new, improved technology that replaces the earlier technology. As product lives shorten in many industries, this latter force increasingly comes to the fore.

At first blush, the decline phase might seem like the least interesting phase of the cycle. The key battles have been fought, the winners have emerged, and there is little or no prospect of additional market growth. Meanwhile, newer technologies are entering into the market and obsoleting much, or all, of the value of the current technologies.

In fact, though, alert managers are crafting business models that employ many profitable ways to extract value from IP that might seem to be of little value. GE exited most of its electronics businesses (televisions, radios, stereos, etc.) a long time ago. Yet one can still readily purchase GE brand products of many kinds, thanks to a clever deal GE did with some Asian manufacturers to supply these goods under the GE nameplate. The manufacturers do all the R&D, manufacturing, marketing, sales, and distribution, and pay GE a royalty for the use of its IP (here, the GE trademark). GE takes no business risk, requires no assets to support this use of its IP, and receives a nice royalty for its name. This is a very profitable business model indeed! And it came from GE's choosing to exit a declining set of businesses that it felt were no longer strategically attractive for the company.

IBM is another company that has made some very profitable lemonade by getting out of businesses it felt had become lemons. When IBM exited the PC business, it found that it had some tremendous IP assets that it could leverage against the leaders in the PC business, such as Dell and Linova. In the case of Dell, IBM packaged its IP assets together with a long-term deal to supply various IT services to Dell and announced a relationship that will result in many tens of millions of dollars being paid to IBM. In the case of Linova, IBM also receives more than $1.7 billion in payment, plus expanded access in China for IBM's IT services businesses. By closing one door (its PC business), IBM has opened another door (its IT services offerings in China). A third exit came in the switches and router business, where IBM was competing against Cisco Systems without much success. When IBM closed its business, it licensed some of its IP to Cisco (again bundled into an announced agreement to sell IT services to Cisco) and received many tens of millions of dollars.

These examples illustrate the value of alert IP management in the declining phase of the IP life cycle model. One factor that adds to the opportunity for value realization in this stage is the fact that the firm is no longer going to be operating in that business. While the firm was in that business, its own assets and operations were at risk of violating another company's IP rights. If the firm elected to assert its IP claims over another firm, that other firm often chose to file a counterclaim against the first firm. That counterclaim alleged that the first firm was infringing on the second firm's IP. The usual course for such issues is a good deal of posturing by both sides, followed by a settlement whereby each side cross-licenses its IP to the other, perhaps with a payment to make up any perceived difference. This was the situation in figure 4-2.

Once the first firm chooses to leave the business, the situation changes. Now the firm no longer has its own assets exposed to the potential counterclaims of the other party. The apparent weakness (namely, the decision to leave the business) creates a stronger position from which to negotiate with others that might infringe on the firm's IP (which, after all, remains in force even after the firm leaves the business). This is the situation in figure 4-3.[15]

Another example of finding IP value in the declining phase of the TLC comes from failed start-up ventures. From 1998 to early 2001, the business world witnessed a boom in start-ups (during the so-called bubble years) that was followed by a corresponding bust. Ventures that had achieved pre-money valuations of $100 million or more had to sell off all of their assets at auction to recover a few pennies on the dollar for their investors. These assets included the IP held by the venture. While many of these ventures had no assets of any real value, some of them may have had patents on good technologies and ideas that failed due to other reasons. These IP assets are typically auctioned off with the furniture by the auctioneer, such that alert buyers can obtain legally valid patents for as little as a few thousand dollars each. By comparison, the process of developing and filing for a new patent that eventually issues is somewhere between $15,000 and $50,000. It is often more expensive to analyze the patents carefully than it is to simply purchase them. Established companies are now rounding out their IP portfolios by shopping at these auctions to acquire additional patents.

Managing Across Phases of the Technology Life Cycle: Windows in China

With the IP life cycle model, the prevailing conception of managing intellectual property with a one-size-fits-all mentality is swept aside. In its place, a more nuanced, dynamic, and strategically focused approach will arise, causing firms to search actively for unrealized business opportunities latent within their knowledge base, even as they manage risks that emerge in their current businesses. While I have already illustrated examples of differing IP management strategies within each phase of the TLC, much of the power of this approach comes from looking across phases of the cycle.

In the earliest phase of the technology, it pays to be very open. Neither you nor others know yet the best use of a particular technology, and no one has an appropriate business model to commercialize any applications either. As the dominant design emerges, tightening the protection for one's ideas becomes very important. In the mature phase, IP management must become more differentiated and segmented, to support different applications of the technology in different uses. In the decline phase, firms can aggressively harvest the fruits of their earlier investments in IP protection.

To see the benefit of this approach across phases of the TLC, consider the problem Microsoft is facing with pirated copies of Windows in China. In the United States and Europe, Windows has become the dominant PC operating system, and growth in those regions now is quite flat—placing it squarely in the mature phase of the technology life cycle. In China, however, matters are quite different. The rising economic prosperity of the country has created a recent boom in the number of PCs selling in the country such that, in this region, the technology life cycle is transitioning from emerging to growth.

One-size-fits-all thinking would suggest that Microsoft should seek to employ the same protections against software piracy in China that it uses in the United States. This would mean that the company (perhaps in concert with other prominent software companies) should vigorously police the use of its software and undertake prompt legal action against any and all illegal use, wherever in the world such illegal activities are found.

A more nuanced view of where Windows falls in the TLC suggests a dramatically different approach. In the United States, Microsoft has won the battle for the desktop. Its Windows operating system enjoys a market share in excess of 90 percent.[16] Even the rival operating system of Linux, from the open source community, poses little real threat to Microsoft's position on the desktop. In China, however, the battle for the desktop is still very much in progress. While Microsoft is in the lead, Linux is making a strong challenge. In fact, the Linux community has signed a deal with the Chinese government to make Linux the default operating system for computers in the Chinese government and many parts of the Chinese educational system.

In this context, applying western IP enforcement policies to stem the flood of illegal copies of Windows in China risks winning the battle (to deter and punish IP infringement) while losing the war (to become the dominant standard on the desktop). So long as Linux remains a serious rival for the desktop operating system of choice in China, Microsoft should actually *welcome* pirated copies of its software. Illegal copies of Windows are free, which helps Microsoft offset the initial cost advantage of "free" open source software. Every pirated copy installed on a Chinese computer that is used by one or more Chinese citizens brings one more person into the Microsoft ecosystem. This strengthens Microsoft's market for third-party developers of applications, tools, and other complementary products (some of which are made by Microsoft itself, so Microsoft can make money on pirated versions of Windows through these other products). More importantly, this approach denies Linux that next new customer that would similarly strengthen *its* ecosystem against Windows.

If Microsoft succeeds in discouraging piracy of Windows in China during this current stage, it is far more likely to drive the user of the pirated software into the Linux camp than it is to drive that person into the ranks of paid-up users of Windows in China. So long as Linux remains a strategic threat to Windows, this is the exact opposite of what Microsoft's IP management should be trying to accomplish. Microsoft's IP management strategy in China should be focused on securing the victory of Windows on the desktops of all PCs in China. That may require deliberately lax enforcement efforts against pirated copies of Windows for the short and medium term. And there are hints that Microsoft may be doing just that.[17] Only after the Linux threat has been eliminated does Microsoft have the luxury of then tightening up the protection of Windows against piracy, as it is now doing in the West.

So Microsoft would be well advised to take a very different approach to managing its IP around Windows in the earlier phase of the TLC in China, when compared to its approach to protecting its IP around Windows in the United States. This suggests that IP management must be driven first and foremost by the business objectives of the company, and not by a legal perspective.

A legal perspective might well be, for example, that lax enforcement of IP in one region sets a bad precedent for IP enforcement in other regions. Another legal view might be to pick an especially egregious case of piracy of Windows in China to set an example for others who might seek to copy Windows illegally themselves. These are worthy points to consider but are subordinate to the strategic objective of establishing the market position of the default standard operating system in China. Unless the legal team is included in all of the key business and strategic decision making for Windows in China, the legal specialists, doing their job as best as they know how, might inadvertently sabotage the overall strategy.

AN OPEN APPROACH TO MANAGING IP

Patents allow you to exclude others from practicing a technology that is covered by your patent. It may not, however, allow you to practice your own technology, if someone else holds patents that read on your approach. This subtle distinction gives rise to a variety of circumstances that the firm must manage as it creates and manages a business model to profit from a technology. In some cases, the firm may achieve freedom of action by cross-licensing. In other cases, though, the owners of other patents may employ business models that make cross-licensing unattractive to them. In this latter instance, cross-licensing will not provide any protection to the business model.

Patent mapping can help identify the risks, and the opportunities, that exist in the value chains in which your business model operates. Hazardous areas can be proactively flagged for specific attention. Here is where you can take steps to protect yourself against the trolls. Areas of opportunity may help direct the entry into related products and services that benefit from the IP portfolio of the firm, enhance relationships with suppliers and customers, or generate new revenue streams.

In the previous chapter, we discussed the emergence of secondary markets for innovations and their associated IP. These secondary markets become a shopping center for IP to help provide protection when entering into new markets. As we saw in this chapter, they can also help create monetary value from IP associated with businesses in which you are no longer operating.

Overall, on both the buy side and the sell side, managing IP requires a phased approach that corresponds to the technology life cycle of the industry. Effective management will vary by the phase in this cycle, instead of using a one-size-fits-all approach. And a truly dynamic approach will harness the management of IP to the larger business strategy objectives, whether those objectives are operating in the United States or around the globe.

5

A Framework for Advancing Your Business Model

In the last chapter, we saw how stronger protection for patents and other intellectual property, combined with the newly emerging secondary markets for innovations and their associated IP, have changed the landscape for crafting business models. In this chapter, we will look at business models themselves in far more detail. The core idea of this book is that companies must develop more open business models if they are to make the most of the opportunities offered by Open Innovation.

Business models are essential for converting ideas and technologies into economic value. However, business models are not all the same. Some confer little or no advantage relative to other companies. Others differ in the extent to which they segment their approach to their target markets. Many business models are closed, making little use of external ideas and technologies. Some are becoming more open, incorporating useful ideas from many places outside the four walls of the company. A few have been able to position themselves as a platform and entice many other companies to make investments and build business models on top of that platform. In this chapter, we will examine a systematic framework for evaluating and then improving your business model.

WHAT IS A BUSINESS MODEL?

The *business model* is a useful framework to link ideas and technologies to economic outcomes. While this term is usually applied in the context of entrepreneurial firms, it also has value in understanding how companies of all sizes can convert technological potential into economic value. In other words, every company has a business model, whether that model is articulated or not.

At its heart, a business model performs two important functions: value creation and value capture. First, it defines a series of activities that will yield a new product or service in such a way that there is net value created throughout the various activities. Second, it captures value from a portion of those activities for the firm developing the model. (For more on what a business model does, see "Functions of a Business Model.")

Business models not only must be developed; they also must be managed once they are developed. Managing business models is an activity that is inherently risky and uncertain. There are many potential ways to commercialize an idea or technology, many of which are unlikely to succeed. And successful business models pose additional risks. They create a strong inertia inside the company that makes any change to the business model much more difficult for the company to accomplish.

Nonetheless, some companies have been able to foster change in their business models. The pioneering firms have done so by responding to dramatic shifts in their markets, often in response to a crisis. Many more have done so by benchmarking the best practices of other companies and mimicking a subset of these practices within their own firms. This has given companies some sense of how to improve their business model, where to improve, and that business model improvement is indeed possible.

But companies need more than a sense that change is possible: they also need a road map that can provide some overall direction for how they might change their business model and how to sequence the necessary changes. This chapter describes a framework that explains the different kinds of business models. The chapter also sketches out the associated innovation processes and IP management that supports each business model.

FUNCTIONS OF A BUSINESS MODEL

A business model encompasses these six functions:

1. Articulate the *value proposition*—that is, the value created for users by the offering

2. Identify a *market segment*—that is, the users to whom the offering and its purpose are useful

3. Define the structure of the *value chain* required by the firm to create and distribute the offering, and determine the complementary assets needed to support the firm's position in this chain (this includes the firm's suppliers and customers and should extend from raw materials to the final customer)

4. Specify the revenue generation mechanisms for the firm, and estimate the *cost structure* and *profit potential* of producing the offering, given the value proposition and value-chain structure chosen

5. Describe the position of the firm within the *value network* (also referred to as the "ecosystem"), linking suppliers and customers, including identification of potential complementors (third-party software developers) and competitors

6. Formulate the *competitive strategy* by which the innovating firm will gain and hold an advantage over rivals

Henry Chesbrough, *Open Innovation: The New Imperative for Creating and Profiting from Technology* (Boston: Harvard Business School Press, 2003).

Business models are not all the same. One way to consider different business models is called the business model framework (BMF). It posits six types of business models—from very basic models with little advantage, to highly sophisticated models that enjoy tremendous advantages. These models vary on two dimensions: first, the depth of investment made to support the business model and, second, the openness of the business model. Like all models, this framework is a considerable simplification of reality. But good models isolate important issues that aren't clearly visible in the immense complexity that accompanies reality. The model seeks to help companies assess where their current business model stands in relation to its potential, and to define appropriate next steps for the further advancement of that model. It also identifies connections between the business model activities of the company, the corresponding innovative activities of the R&D organization, and the management of IP associated with the business model.

THE BUSINESS MODEL FRAMEWORK

Here are the six kinds of business models:

Type 1—Company has an undifferentiated business model

Type 2—Company has some differentiation in its business model

Type 3—Company develops a segmented business model

Type 4—Company has an externally aware business model

Type 5—Company integrates its innovation process with its business model

Type 6—Company's business model is able to change, and is changed by, the market[1]

There are innovation processes and IP management aspects associated with each type of the model. See table 5-1 for the matrix that describes these interrelated aspects.

The sections that follow will illustrate each type of the framework and the associated innovation and IP management implied by that type. To be more concrete, some examples are provided that help illustrate companies' processes for each type of model. Let's start at the beginning, or type 1.

TABLE 5-1

The matrix of the business model framework, with its associated innovation and IP management processes

	Business model	Innovation process	IP management
Type 1	Undifferentiated	None	NA
Type 2	Differentiated	Ad hoc	Reactive
Type 3	Segmented	Planned	Defensive
Type 4	Externally aware	Externally supportive	Enabling asset
Type 5	Integrated	Connected to business model	Financial asset
Type 6	Adaptive	Identifies new business models	Strategic asset

Type 1: Company Has an Undifferentiated Business Model

Every company has a business model, whether that model is articulated or not. While definitions of business models vary to some degree, experts agree that business models are the intended ways that companies will make money out of their ideas, resources, and technologies.[2] However, the vast majority of companies operating today do not articulate a distinct business model and lack a process for managing one. These companies are operating with type 1 business models. A business using the undifferentiated model competes on price and availability and serves customers who buy on those criteria. In other words, firms using type 1 business models are selling commodities and are doing so in ways that are no different from many, many other firms. They often are caught in the "commodity trap."

Companies operating in type 1 find it enormously difficult to sustain any competitive advantage in their business. Type 1 companies sometimes do change, either by copying an idea they observe in another company or hiring someone from another company who can teach them something new. But they are more likely to fail to change and be swept away as improvements come into their industry, but not into their company.

Since these companies rely extensively on copying others, they seldom, if ever, get new innovations implemented first. And any advantage that comes their way is equally difficult to protect from someone else copying it. What this means is that a type 1 company lacks the ability to control its destiny. When a superior technology comes into its industry, the type 1 company lacks the business model to respond. When the market served by a type 1 company saturates and disappears, the type 1 company is likely to disappear too.

Companies with type 1 business models also lack much of a process to innovate and manage IP. About 15 percent of U.S. employees worked in companies conducting R&D, and about 11 percent of U.S. employees worked in companies that spent more than $1,000 per year on R&D.[3] In other words, most U.S. employees work in firms that do not possess much of an innovation process. These are the firms that are not likely to generate any significant IP.

Not only do type 1 companies create very little IP; they also lack the resources to defend what little IP they have created. While we hear stories in the press of small inventors winning large infringement cases against large corporations, these are very much the exceptional cases. More typically, type 1 companies must soldier on with little to protect, and little with which to protect it. Most inventors have limited success against larger, better funded, more established organizations.

To be fair, there are advantages to the type 1 business model. First and foremost, the type 1 model is by far the lowest-cost model. This model reduces the cost of a firm entering into a new market since it requires no money to be spent on expensive items such as innovation. It is also a useful starting point for entrepreneurs who are seeking employment for themselves, their family members, and perhaps some of their friends. Hard work and good fortune can carry a type 1 company some distance. But such a company will find it hard to attract investment capital and to scale up its activities. If the type 1 company seeks to endure, it will need to move beyond its origins and advance its business model to the next type.

Examples of type 1 companies include mom-and-pop restaurants; many family farms; many independent bookstores, cafés, and barber shops; and, indeed, many kinds of entry-level services establishments. In semiconductors, type 1 companies include developers and purveyors of commodity products. In pharmaceuticals, examples include certain contract research organizations or generic drug manufacturers. In entertainment, examples include the myriad labels out there trying to sign artists, represent up-and-coming talent, or develop new concepts for movie scripts. (Other companies that you may think are type 1 have much more complicated business models. See "What About Dell and Wal-Mart?")

Type 2: Company Has Some Differentiation in Its Business Model

Other companies—type 2 companies—create some degree of differentiation in their products or services. This leads them to differentiate their business model as well, allowing them to target customers other than those that buy simply on price and availability (such as a performance-oriented customer). This allows the type 2 company to serve a different and less congested market segment from that served by its type 1 counterpart. The ability to differentiate itself from its numerous type 1 competitors supports a period of growth for the company. If the differentiation is high enough, the company may also enjoy a period of above-normal profits.

The type 2 firm has escaped—at least for a while—the commodity trap that imperils the type 1 firm. While there is often innovation activity in the type 2 firm (which is one source of the differentiation it enjoys), this innovation is ad hoc in its nature. It is not well planned, and budgets are dictated by what can be afforded, not by what is required. Often, there is little organized process and insufficient funds to develop further innovations. The company's primary focus is on the execution of its business, and the type 2 company may lack the resources and staying power to invest in the supporting innovations to sustain its differentiated position. For example, a performance advantage can only be sustained through the generation of future performance enhancements, since any current advantage will eventually subside as others copy, catch up, or perhaps overtake that enhancement.

WHAT ABOUT DELL AND WAL-MART?

Thoughtful readers might raise a powerful objection at this point. The type 1 company does not invest much in innovation and therefore faces an uncertain future. What about companies like Dell and Wal-Mart? They are well known for investing very little of their own funds in innovation, yet their future seems assured for many years to come.

It is true that companies like Dell and Wal-Mart do not invest much of their own money in product innovation. Each company, however, is a leader within its industry on *process* innovation, which is at least as important to future success as product innovation. For example, Dell has pioneered the ability to order a wide variety of PCs direct from the manufacturer and has invested significant sums in creating the processes that can accept and fulfill orders in just forty-eight hours. Wal-Mart has led the way in electronic document interchange bar coding with its suppliers and is today the leader in getting its supply chain to adopt radio frequency identification (RFID) devices. These are significant process technology advances.

Both companies also drive innovation throughout their supply chain. They enjoy such enormous economies of scale that they can induce their supply chain to build business models around serving them. Suppliers now make investments that include warehouses at the sites of Dell or Wal-Mart, carrying extra inventory, adopting technical standards, and, of course, selling to these retailers at the lowest prices (as we will see, the ability to induce these investments is an attribute of companies with very advanced business models in the BMF). These and many other process innovations give each company a powerful, sustainable advantage in its markets.

This type is characteristic of many start-up companies that are pushing a new technology in the emerging phase of the TLC we examined in the previous chapter. They do possess some differentiation in their technology and are engaged in a contest to become the dominant design.

Dell and Wal-Mart also innovate in another important way. Each company must decide whether and when to add new kinds of products and services to its lineup. This is a vital capability for each company's future, because every product has a life cycle and will eventually drop out of the market. To grow, these companies must stretch to embrace new kinds of items within their business models. In some cases, like RFID, they also must innovate their business models to work with new technology.

These decisions of whether and when to add new products are risky. If Dell and Wal-Mart wait too long, other companies will move ahead of them and take away market share. If they move too early, their internal systems will incur a great deal of cost, with insufficient revenue earned to cover those costs. Dell, for instance, has expanded its business from PCs and notebooks to printers and servers as well. This is fortunate because the PC and notebook markets have matured, while the newer segments are still growing. While many thought that Dell lacked the technical skill to sell server products, the company has forged partnerships with Intel and EMC and is rapidly gaining share in this more advanced, more sophisticated market. Dell is also developing better ways to collaborate with key technology partners, a process that will confer future advantages as the company seeks to enter still more segments.

If we consider the combination of process technology investments, the challenges of adding new and often more complex products to their businesses, and the ability to make suppliers construct business models around their needs, it becomes clear that Dell and Wal-Mart do not employ type 1 business models.

There is some level of organization to the innovation process in a type 2 company. The innovation activity in the type 2 firm is typically led by the CEO or one of the key technical leaders. Some IP is developed within the firm, and some resources are dedicated to creating and defending the IP. This is usually done by working with outside patent counsel. However, IP management is rather reactive and haphazard in a type 2 company. Because IP is created only occasionally, it is not something that is planned and managed on a regular basis.

This type of innovation maturity is typical of many young companies, especially technology-based start-up companies. Tech start-ups are given one round of funding and must demonstrate the viability of their technology and develop a business model within the constraints of that funding. Unless they are extraordinarily successful, they lack the funds to innovate beyond extensions of the first concept. This type is also typical of many individual inventors. These inventors come up with a successful invention that they are able to license or commercialize. But they lack any capability to follow up on this achievement. This situation gives rise to the pattern of so-called one-hit wonders, where a company or inventor has a successful first product but is unable to follow up this success with additional products of similar success.[4]

There are many examples of these one-hit wonders. GO and Collabra, from chapter 2, fit this profile. In the hard disk drive industry where I used to work, many newly entering companies differentiated their products on the basis of providing higher performance than the industry average. Most such firms were out of business within a few years, owing to the inability to come up with a second generation product that enjoyed a similar performance advantage. Semiconductors is another industry with many one-product companies. In pharmaceuticals, many university spin-off companies start out with a strong technology advantage within a certain niche. They then sign a deal with a large pharmaceutical firm to commercialize that product, only to discover that they lack an adequate follow-on technology to launch a new product. (One-hit wonders are not confined to the business world. In entertainment, many radio stations play hit songs from groups that never were able to come up with any additional hit songs. There are even Web sites devoted to the phenomenon.[5])

The problem with being a one-hit wonder is that a type 2 company cannot sustain its success beyond its initial products or services. There isn't enough depth of investment to support and sustain the business model. This creates a pattern of a rapid rise, due to the success of the innovation, only to be followed by rapid descent, as the innovation becomes increasingly obsolete and no new innovation exists to carry the company on from there.

To summarize, there are key differences between a type 2 company and a type 1 company: (1) some differentiation is achieved by the company through its business model; (2) there is now innovative work being done; (3) some IP is being generated and occasionally defended.

Type 3: Company Develops a Segmented Business Model

A lot of work has to happen before a company moves its business model from type 2 to type 3. Now there is much greater depth of investment needed to support the business model. Type 3 companies are able to do more planning about their future in part because they have developed a business model that allows the company to begin to segment its markets. The company now can compete in different segments simultaneously. More of the market is thus served, and more profit is extracted from the market as well. The price-sensitive segment provides the volume base for high-volume, low-cost production. The performance segment supplies high margins for the business. Other niches can also be addressed, creating a stronger presence in the distribution channels.

To return to the TLC model of the previous chapter, companies that win the battle to become the dominant design often find themselves in this business model type. The growth in the market that has come from being the dominant design is fueling the ability to segment the market and is providing the resources to develop offerings for multiple market segments.

The firm's business model now is more distinctive and profitable, which supports the firm's ability to plan for its future. The company relies on its business model to select useful outcomes from its internal R&D activities and commercializes those outcomes through its business model. Periodically, the new projects advance the company's business.[6] The ability to plan also enables the company to look a little further out in its innovation activities (up to perhaps three years), and to seek out new segments of the market that it could serve in the future.

In a type 3 company, innovation is no longer a random occurrence. Rather, innovation becomes a planned activity, with ongoing financial and organizational resources committed to it. There often is a department dedicated to pursuing innovation (usually the engineering or R&D department). This greater level of planning and resource commitment helps the type 3 firm avoid some of the hazards of the one-hit wonders found in type 2 companies.

One important indicator of the more planned and organized nature of innovation in type 3 companies is the creation of road maps of future products and services. By a *road map*, I mean a look at what specific products and services are intended to be offered by the company at specific dates in the near future (usually one to three years from the present). Unlike a type 2 company's road map, the planning now encompasses multiple segments. Not only is there a road map, though; this road map is also supported by schedules and budgets that will enable the organization to meet the plans. Innovation budgets now begin to reflect the requirements of sustaining the business, not just what can be expended in the current period for innovation.

Because of this more planned and further-ranging R&D activity, the firm begins to build its IP portfolio. In a type 3, the function of IP management starts to become a full-time activity within the organization (instead of relying primarily on outside patent counsel). Type 3 companies begin to use patent maps to identify areas of vulnerability in their value chain, as we discussed in the previous chapter.

Organizationally, innovation in a type 3 company involves multiple functional areas. While the CEO reviews innovation activities, primary responsibility now falls on one of the CEO's direct reports, usually the head of engineering or R&D. Engineering is usually the driving area, as in type 2, but now engineering solicits customer input through the sales organization, while supplier input is actively sought through the purchasing organization. The road maps help engineering coordinate critical interactions with the other functional groups in the company.

While its greater level of planning helps the type 3 company avert the one-hit wonder syndrome, problems still remain. Type 3 companies think of innovation from a product or technology perspective. While they are alert to opportunities within the boundaries of the current business and market, they do not see innovation as being able to stretch those boundaries. The type 3 firm remains vulnerable to any major new technological shift beyond the scope of its current business and innovation activities, and also to major shifts in the market.

Examples of type 3 companies include companies with good product or process technologies. These might include young start-up companies that have grown beyond the one-hit wonder risks of type 2 by creating additional sets of product or process technologies. Examples also include many industrial age companies that have built a well-earned reputation for prowess in a particular product or technology and are now struggling to adapt their processes to the requirements of the digital information age.

One of the subtle challenges facing a type 3 company is that it thinks of innovation as product innovation or process innovation, without considering the business dimensions of innovation. Because the business model is taken for granted, its influence on what ideas are considered and what ideas are rejected is usually not examined sufficiently. This was a large problem for Xerox with the many projects that spun out of its Palo Alto Research Center. Projects that did not fit with the "razor and razor blade" business model of copiers and printers (where you sell the razor at a competitive price and then earn most of your profits from the razor blade that your razor requires) created tremendous value when placed into ventures that used very different business models in the nascent computer industry.[7]

To summarize, there are a number of key differences that separate a type 3 company from type 2: (1) the company segments its markets and serves multiple segments, and it selects innovation projects from among a number of possible projects based on its business model; (2) innovation is a planned organizational process, not a random event; (3) innovation is treated as an investment in the company's future, enabling the company to look into its future; (4) functions beyond engineering or R&D are a part of the innovation process; and (5) IP management is managed as someone's responsibility inside the firm.

Type 4: Company Has an Externally Aware Business Model

In this business model, the company has started to open itself to external ideas and technologies in the development and execution of the business—an important distinction from the type 3 business model. This unlocks a significantly greater set of resources available to a company. This greater variety of resources helps the firm search more widely for possibilities.

With a type 4 business model, the company continues to segment its markets through its business model, supporting the segments with its innovation process. Now, however, the segmentation is supported by external sources of technology, as well as internal sources of technology. Type 4 business models cross a conceptually important threshold that was not part of the earlier business models; they signify the beginning type of more open business models to come.

The type 4 business model is externally aware and selectively incorporates external innovation inputs into the business. Such external innovation reduces the cost of serving the business, reduces the time it takes to get new offerings to market, and shares the risks of new products and processes with other parties.[8] This outside innovation extends the range of segments able to be addressed by the business model. It enables the company to serve customers not only when the company produces the product or service but also when the company integrates the outside item into its offering.

The business model not only shapes the selection of internal innovation projects, it now begins to shape the cultivation of external innovation opportunities as well. The road maps of the firm provide a shopping list of needs within the firm for external ideas and technologies. Relationships with outsiders help identify external projects that might fulfill some of these needs.

The business model also begins to act as a source for new growth for the type 4 company. Adjacent markets become areas of investigation for extending the firm's business model. Growth now can come from penetrating the current market further and/or from applying the business model to adjacent markets that can be served with that model.

This type of business model corresponds nicely to the mature phase of the TLC. In the mature phase, there are many opportunities to apply the dominant design in new market segments. Companies typically lack the resources to pursue most of these new niche opportunities. Opening up one's technology to others, and allowing them to build on it to serve new markets, is one way to serve new segments without losing focus in the main market.

In addition to the planning and organizational commitments of type 3, innovation is now driving the company to look outside for ideas and inputs to the innovation process. This more external perspective manifests itself in manifold ways. One is that internal road maps are now shared with suppliers and customers on a frequent basis. This enables the firm to make much more systematic use of innovative ideas from suppliers and from customers. It also allows suppliers and customers to plan their own activities in concert with the innovative activities of the firm. Another role that emerges is an active technical advisory board, populated by technical and industry experts. This TAB provides a forum for external input to come into the firm, well beyond the inputs available from suppliers and customers. Universities, for example, are now contacted and cultivated for possible new ideas. Often, formal projects and ongoing linkages are forged out of TAB connections.

The perspective of the type 4 company toward innovation begins to shift from a product/process/technology focus toward a business focus. This further reduces the one-hit wonder problems of type 2 and begins to address the short-term myopia that is still present in type 3. The type 4 firm can now initiate change in related areas, rather than react to change initiated by others. However, the type 4 firm still focuses its R&D activities on current and adjacent areas. It therefore remains unprotected from innovations that arise in seemingly unrelated areas and invade its markets.

Organizationally, innovation becomes a cross-functional activity. The marketing function is now involved in innovation as an equal partner with engineering or R&D. Instead of relying on sales inputs for new projects, a more systematic investigation of current and potential customers, in current and potential markets, is part of the innovation process. Complete business cases are developed for alternative innovation opportunities, with the finance organization playing an important role in developing and vetting these cases. This enables the type 4 firm to plan further out in time, and to evaluate the financial risks of longer-term R&D investments within a portfolio of projects.

IP management becomes a business function in type 4 companies, with its own financial and organizational objectives. IP is viewed as another class of corporate assets. The costs of creating and protecting IP are weighed against the benefits of having that IP. Budgets for IP are created, with employees tasked with the management of that budget. IP mapping is now commonly done, and external IP may be bought or licensed to help strengthen vulnerable elements of the value chain. Unused internal technologies are occasionally licensed outside for additional revenue.

Examples of type 4 companies include many industrial firms with established corporate R&D activities. Companies like RealNetworks that are actively working with external technology as well as their own technology fit here. Many drug companies that are beginning to work more closely with start-up biotech firms and university spin-off companies also fit here. Food companies are also moving in this direction.[9] Some banks and financial institutions are also type 4, as they reach out more systematically to their customers and markets for innovative ideas. Younger technology companies that have managed to navigate transitions from one technology to another by partnering with outside firms may also fall into this category.

A recent example of a powerful type 4 company is the German software company SAP, with its market-leading business process integration software, R/3. R/3's situation indicates both the strength and the potential weakness of the type 4 firm. On the one hand, R/3 leads its market segment in sales and is highly profitable for SAP. On the other hand, R/3 is a deeply vertically integrated technology, built with proprietary tools, based on SAP's proprietary software programming language and its proprietary knowledge of its customers' business processes. For more than a decade, SAP has enjoyed tremendous growth with R/3. Its main strategy for openness was its distribution: it partnered closely with IT consulting firms that performed the configuration and installation tasks necessary to run the software. Now, however, new technologies are coming into the IT market that offer widespread connectivity across business processes, such as service-oriented architectures (SOAs). R/3 risks becoming isolated, and its customers risk being marooned on a proprietary platform while the world moves toward a more open, interconnected architecture. R/3's business model is thus at some significant risk, despite its great success to date.

To summarize, there are again several key differences between a type 4 business model and a type 3: (1) the type 4 business model incorporates external technologies in serving current customers and can be extended to adjacent markets for new growth; (2) the type 4 company begins to look outside proactively for innovations as well as inside; (3) there is a far greater role for suppliers and customers in the innovation process; (4) innovation becomes a cross-functional activity between multiple internal functions that have equal standing; and (5) IP is now managed as an enabling asset, helping access adjacent markets and generate value.

Type 5: Company Integrates Its Innovation Process with Its Business Model

In a type 5 model, some very exciting things happen to the business model of the firm. The company's business model now plays a key integrative role within the company. There is a strong shared sense of the business model—both what it can do and what it cannot do—that connects the many different functions of the company together and helps them work through complex challenges effectively. This shared perspective on the business model extends outside the firm as well, with external parties that understand the kinds of innovations that the firm is looking for.

Suppliers and customers enjoy formalized institutional access to the firm's innovation process, and this access is reciprocated by the suppliers and customers. For example, company personnel may be invited to join the TAB of one or more suppliers, and company personnel receive regular briefings on suppliers' road maps. Customers share their own road maps with the company, giving the company much better visibility into the customers' future requirements. And customers and key suppliers are engaged at multiple organizational and functional levels for innovation insights.

Nor does the visibility into the supply chain or customer base stop there. Type 5 companies take the time to understand the supply chain all the way back to the basic raw materials, as they look for major technical shifts or cost reduction opportunities. Type 5 companies also invest substantial resources to study "the customer's customer" to learn about the deeper unmet needs and opportunities in the market. Distribution channels are in play as well. While current channels are leveraged as much as is practicable, alternative distribution arrangements are actively considered. Some experimentation is conducted on alternative distribution channels and, indeed, on alternative configurations of the business model.

The type 5 business model uses its understanding of customers and suppliers to identify discrepancies and disconnections between the customer's or supplier's business model and the company's own business model, both in the current business and in new business areas. These issues are proactively identified, and actions are taken to address the situation, so that the company maintains alignment of its business model with that of its customers and key suppliers.

This type also fits well with the mature phase of the TLC. Now the company is able to forge strong alliances and partnerships with complementing firms, as they tap the market opportunities in new market areas. Because the investments and risks are shared with its partners, the company with its partners can search and serve a wider market space at lower cost. The company not only can offer its business model to external parties to incorporate their technology; the company can also offer its technologies to the business models of external parties.

Innovation is becoming embedded inside the corporate DNA in companies operating with a type 5 business model. Staff members from every functional area feel that they are able to contribute to the future of the company. In type 5, the company has become an effective integrator of both internal R&D and external R&D. The business model consciously considers how to create systems and architectures that make the best out of both, and begins to conceive of itself as a platform to connect and coordinate innovative activities.

IP management takes on a more strategic character as well. Patent mapping identifies revenue generation opportunities, as well as risk reduction opportunities. External technologies are actively sought in the secondary market for strengthening the internal IP portfolio. External licensing is set up as a profit center, with quarterly and annual targets for revenue, and budgets allocated to support licensing as a business. In short, the firm begins to manage IP as a financial asset, seeking how best to optimize its value on both the sell side (with the firm's own underutilized technologies) and the buy side (searching for external technologies). The firm has begun to track these second-ary markets, with an internal team and external intermediaries both contributing to its efforts.

Organizationally, innovation is viewed in a type 5 company as a business function, led by a senior manager. Engineering, marketing, and finance collaborate in developing and managing the business model through cross-functional innovation teams. The earlier product/technology perspective is now replaced by a business focus on innovation. This places the need to identify and understand shifts in markets and customer needs on par with knowledge of technical changes and opportunities.

Examples of companies with type 5 business models include those that have embraced external sources of technology and are ac-tively building business models that build on those technologies. One telltale sign of this phenomenon comes from the creation of processes to actively promote the firm as a partner of choice for others to approach with innovation opportunities. A number of firms have awakened to the value of branding themselves in this way. In IT, IBM Global Services has tried to foster a reputation of the partner of choice for its IT customers, as well as with new start-up firms.[10] In drugs, Eli Lilly has undertaken a very public initiative to recruit young biotech companies with promising compounds to come work with it. In toys, the Big Idea Group, which we will examine further in the next chapter, strives to be the inventor's partner of choice. In consumer package goods, P&G is out in front, promoting itself as the leader in Open Innovation, even as Kraft, MasterFoods, and other consumer products companies start up their own Open Innovation activities to increase their intake of external ideas and technologies.

A further example shows how a company can move from one type of the framework to a more open type. There is a new initiative from SAP that explicitly addresses the risks faced by its market-leading R/3 product, which were discussed in the previous type of the BMF. This is called Enterprise Service Architecture by SAP, and it seeks to connect the R/3 business processes with the Internet standards of SOA, including the .NET and WebSphere architectures of Micro-soft and IBM. SAP is even marketing a new product, NetWeaver, that will help customers and third-party software companies connect R/3 and other SAP products to these wider architectures. SAP is beginning to think of itself differently now. Instead of being the leading business applications software provider (and furnishing all of those applications itself in a tightly integrated manner), it now sees itself as delivering tools for connecting business processes developed both inside *and* outside the company. It realizes that not all of these applications to connect business processes can or should be provided by SAP. If SAP can execute this transition successfully, its business model will move from type 4 to type 5.

To summarize, there are several key differences in moving the business model from type 4 to type 5: (1) the company's business model is now focused on new markets and new businesses, as well as current business, and the company strives to align customers and suppliers with its business model; (2) the company's internal and external R&D activities are integrated through the company's widely understood business model; (3) the company's innovation road maps are widely shared with suppliers and customers, and this access is reciprocated by those parties; (4) innovation is a business function, in which functional heads are led by a senior business manager; and (5) IP is managed as another kind of financial asset, managed within a profit center.

Type 6: Company's Business Model Is Able to Change, and Is Changed by, the Market

The type 6 business model is an even more open and adaptive model than types 4 or 5. One important attribute of a company with a type 6 business model is its ability to innovate its own business model. This requires a commitment to experimentation with one or more business model variants, and a willingness to invest some amount of funds and management attention to explore alternative ways to profit from innovation. This experimentation can take a number of different forms. Some companies use corporate venture capital as a means to explore alternative business models in small start-up companies. Some use spin-offs and joint ventures as a means to commercialize technologies outside their own current business model. (Later success with a spin-off or joint venture, in turn, might help the company shift its own business model in that direction.) Some have created internal incubators to cultivate promising ideas that are not yet ready for high-volume commercialization.

Experimentation with the business model also extends to customers and suppliers. In type 6, key suppliers and customers become business partners, entering into relationships in which both technical and business risks are shared. The business models of suppliers are now integrated into the planning processes of the company. The company in turn has integrated its business model into the business model of its key customers. This allows the company to create its business model as a platform to lead its industry, including suppliers and customers. And this platform effectively organizes and coordinates the work of many others in the service of the business model.

Dell is a good example of both the supplier and customer dimensions of this type 6 partnership with suppliers and customers. With regard to its management of suppliers, Dell segments its suppliers, just as it segments its customers. Its relationship with Intel, for example, goes far beyond that of a traditional supplier or vendor. Dell works closely with Intel on future technology planning. It acts as an early test bed for new Intel chips and often is the first company to develop a new motherboard for the next-generation chip. It shares all of its field failure data with Intel. And, until May 2006, Dell purchased its chips exclusively from Intel, rather than buying chips from Intel's rival, AMD.

Dell also works closely with its enterprise customers, which it segments (and treats quite differently) from its consumer customers. Dell maintains a database of all Dell products sold at each enterprise. The customer can specify a three-year or four-year rotation period for receiving new Dell computer products, and Dell effectively administers the company's rotation policy on behalf of the customer. It will even create and install a customized software configuration for that customer so that each employee at each company location receives a new PC or notebook every three or four years, loaded with the exact software specified by the customer. This saves IT costs for Dell's customers and creates strong incentives for ongoing relationships with Dell, which in turn generate more data for Dell on all of the users at each enterprise customer.

One important device that enables this integration of business models throughout a value chain is the ability of the company to establish its technologies as the basis for a platform of innovation for that value chain. In this way, the company can attract other companies into its business by sharing the tools, standards, IP, and other know-how that are needed for these supporting players to successfully implement the platform. This platform not only coordinates internal R&D with external R&D toward desired business objectives; it also shapes the future direction of that coordination.[11] It further extends coordination beyond the value chain to the surrounding value network or ecosystem in which the investments of third parties add additional value to the platform itself.[12]

This has been the happy achievement for Apple's iPod. The success of the iPod and Apple's business model for the iPod now elicit new types of accessories, and a variety of enhancements that collectively amount to significant industry investment in the iPod platform. Still others are exploring ways to use the iPod for recording and displaying medical information, financial information, and other real-time information. The presence of this wealth of additional complementary items greatly boosts the value of the iPod. Yet, Apple does not pay anything to induce these investments. Others are investing money that will help Apple make more money. Talk about a great business model!

As the iPod moves from the growth phase to the mature phase of the TLC, however, Apple's business model will likely have to change as well. To harness the iPod platform for a variety of new market segments, Apple will have to open up more of the iPod architecture to allies and partners in order to exploit those opportunities. Apple has begun to do this already with cell phones and its iPod technology, but it will have to partner more extensively for medical markets, financial information uses, and so on. Apple will also need to inject new external technology into the iPod architecture to stay ahead of improving MP3 and other rival media playback technologies.

External licensing in type 6 companies has become part of the organizational DNA within the overall innovation model. The NIH syndrome is no longer an issue, and external technology is put on an equal footing with internal technology. Similarly, it is quite natural for these companies to actively outlicense underutilized internal technologies as well. IP is no longer merely a financial asset. It is now managed as a strategic asset, enabling the firm to enter or exit markets, foster spin-offs or spin-ins, build ecosystems within markets, and make money. The central IP management organization exists as a center of excellence to support the business units in their management of IP. The firm engages with the secondary market for IP on a sustained basis and enjoys superior knowledge about what the going rate is for a variety of technologies that it might choose to buy or to sell. It also enjoys a preferred relationship with IP intermediaries and market makers, enabling it to be presented with opportunities well ahead of most other firms.

IP now is managed in a variety of ways beyond revenue generation. Patent mapping is used both to manage risk and to identify potential reward within current and possible future markets. It also helps identify potential new businesses that might leverage the IP of the company, using that IP to offer an entry ticket to new businesses into the company. Alternatively, IP may provide a consolation prize for the company when it is exiting old businesses. IP may help define the means by which risks and rewards are shared with key partners.

This approach to managing IP accords well with the decline phase of the TLC, because it converts the decline phase into a renewal phase for the firm. Instead of simply withdrawing from the business, the company may choose to empower a business partner to take over the business, while earning revenues for its technology and know-how. Or the company might adopt a new and different business model, which reframes the business in a way that allows the company to deploy a new configuration of assets and resources. For example, IP might be used as a strategic asset to enhance relationships with suppliers and customers, and other parts of the value chain. IP may be harnessed to help kick-start a new ecosystem. In some cases, IP may be bundled in as part of the terms of a purchase or a sale of a product or service. In other cases, IP may be used to help set standards within the supply chain or the customer base. In still other cases, IP may create second sources to instill more vigorous competition in the supply chain. This dynamism and agility, combined with the close collaboration with key partners that themselves invest alongside the firm, separate the type 5 and type 6 business models.

To summarize, there are a number of key differences between type 5 and type 6 companies: (1) the type 6 company's business model drives the business models of its key suppliers and customers; (2) innovating the company's business model, which is widely shared across the company, itself is part of the company's innovation task; (3) external partners share technical and financial risks and rewards with the company in the innovation process; (4) IP is managed as a strategic asset, helping the company enter new businesses, align with suppliers and customers, and exit existing businesses; and (5) the management of innovation and IP is embedded in every business unit of the company.

Examples of this type 6 business model can be found in IBM and Procter & Gamble. IBM is teaching the world a great deal about the many different ways to manage patent portfolios, which today include donating patents to an open source commons, operating alongside a highly profitable outlicensing program. Procter & Gamble has led the way both in the development of techniques for seeking out external technologies for its own brands, and also in the use of outlicensing to create value from its unused internal technologies through other companies' brands. We will explore these companies further in chapter 8.

A TOOL FOR ASSESSING THE TYPE
OF YOUR BUSINESS MODEL

Unless you are working in a brand-new start-up company, your company already has a business model. To advance your model to a different type, you must first assess where you currently are in the business model framework. The key issues for improving your business model will depend on the type of your current business model in terms of the depth of investment you make to support your business, the openness of your business model to external ideas and technologies, and your willingness to let other businesses use your own unused ideas.

To help you assess where you currently are, and what issues are the most important for that type, table 5-2 provides a list of assessment questions, organized by each type of the model. This is intended to guide your own assessment; these questions are by no means exhaustive. Even after working through the questions, you will need to exercise your own judgment.

TOWARD A WORLD OF
OPEN BUSINESS MODELS

In the Open Innovation model, it is the firm's business model that drives its search for innovation activities (from internal or external sources). Firms must search for useful technologies that can advance their business model from whatever sources can provide the appropriate opportunities at the right time. They should be much more open about sharing or licensing technologies that don't fit with their business model but might work well in another company's model. They should then construct open business models to sustain these advances.

Taking this open view of the business model also requires a different view toward IP. The open business model view does not examine individual IP assets in isolation, limited to the consideration of specific individual technologies or know-how. Rather, it views IP in bundles or clusters within an overall portfolio that supports the business model. Synergies between individual elements of IP are consciously identified. Internal IP that is not supporting the business model becomes a candidate for external licensing or an outright sale. External IP that complements the business model becomes an attractive candidate for acquisition from the outside.

When ideas and innovations connect directly to a company's business model, they create additional power and leverage for the other parts of the strategy. Conversely, when these linkages are absent, even very good ideas can be worth little or nothing, because they lack the other elements required to turn an idea into real value. As we shall see later in the book, there are specialist firms emerging that have dedicated their own business models to activities that can help you use IP to advance your business model. These include IP intermediaries that help locate IP for purchase or sale. They include

TABLE 5-2

Diagnostic questions for assessing your business model

	Type 1	Type 2	Type 3	Type 4	Type 5	Type 6
Description	Undifferentiated	Differentiated	Segmented	Externally aware	Integrated with business model	Platform player shapes markets
Examples	Mom-and-pop restaurants	Start-up technology companies	Technology push companies	Mature industrial R&D firms	Leading financial firms	Intel, Wal-Mart, Dell
Diagnostic questions	• Is there anything that differentiates this business from its competitors? • Why do customers buy from us? • Why do customers leave us? • What control do we have over the future direction of our business?	• Do we earn a price premium for our product or service? • Can we sustain our differentiation over time? For how long? • Are we likely to develop a second successful offering? When?	• Are we an engineering-driven company? • Have we created new market segments, or did our customers find us? • Can we further segment our markets? • Can we extend our markets?	• Do we look outside regularly for new ideas and technologies? • Do our key customers and suppliers know about our future road maps? • Is marketing an equal partner in the innovation process?	• Is our business model widely understood within our company? • Do our key customers and suppliers share their road maps with us? • Is innovation managed as a business or as a technology function?	• Can we direct the future evolution of our markets? • Will customers and suppliers fit their business models to ours? • Do other companies routinely invest in projects that require our technology as a platform?

How different from previous type			
NA	• There is innovative work being done within the type 2 firm. • Some differentiation is achieved by the company through its innovations and perhaps through its business model. • Some IP is being generated and defended.	• Innovation is a planned organizational process. • Innovation is treated as an investment in the company's future. • The company segments its markets and serves multiple segments. • Functions beyond engineering or R&D are part of the innovation process. • IP management is coordinated inside the firm as someone's responsibility.	• The type 4 company looks outside for innovations. • There is a role for suppliers and customers in the innovation process. • The business model can be extended to adjacent markets for new growth. • Innovation becomes a cross-functional activity. • IP is managed as a corporate asset, with occasional outlicensing of underused internal technologies.	• The company's internal and external R&D activities are integrated through the company's widely understood business model. • The company's innovation road maps are widely shared, and access is reciprocated by those parties. • The company's business model is focused on new markets and new businesses, as well as current business, and the company is able to align its business model with customers and suppliers. • Innovation is a business function. • IP is managed as another kind of financial asset.	• The company's business model is interconnected with the business models of its key suppliers and customers. • Innovating the company's business model itself is part of the company's innovation task. • External partners share technical and financial risks and rewards with the company in the innovation process. • IP is managed as a strategic asset, helping the company enter new businesses and exit existing businesses. • The management of innovation and IP is embedded in every business unit of the company.

merchant banks that will lend against the value of your IP, and aggregators that will help you liquidate your unwanted IP. They include firms that will assert your IP rights against other firms, in return for a percentage of the monies raised. They also include venture capital and private equity firms that specialize in extracting underutilized IP and other assets to form new companies.

Managing internal and external innovations and IP within an open business model requires the construction and support of a rich internal innovation network, connected to a diverse external innovation community. A few companies have made this transition already. For most others it will require more substantial changes to the business model and the structure of corporate management processes. This will require a lot of hard work by a large number of people inside the organization. But the effort will be well worth the time and trouble, for its completion will harness the efforts of many, many more people who do not work within the firm but whose efforts, properly stimulated and subsequently integrated, will sustain the most enduring businesses of the twenty-first century.

6

Innovation Intermediaries

Chapter 1 explored the many opportunities available to companies that wish to open up their innovation process. Chapter 2 explored some of the challenges facing companies that wish to open up their innovation process. Chapter 3 explored the impact of stronger IP protection on the design of business models and discussed the emergence of secondary markets for innovation. Chapter 4 examined how to link the management of IP to one's business model, while chapter 5 presented a six-stage model for advancing the business model.

This rich environment of abundant and widely distributed knowledge characterized by these chapters can be daunting for a firm. How can you know where to seek out information in this turbulent context? How can you manage the IP involved when working with ideas that originated outside your firm? How can you manage IP when letting others use your ideas in their firm? The good news in this chapter is that others have already begun creating businesses, and their own business models, to help you tackle these Open Innovation challenges.

THE DIFFICULTIES OF SEARCHING FOR, AND EVALUATING, EXTERNAL TECHNOLOGIES

As we discussed in chapter 3, there are real difficulties that companies encounter when they wish to seek out external technologies for their business—or external markets for their own technologies. This is the Arrow Information Paradox: "I as a customer need to know what your technology can do, in quite some detail, before I am willing to buy it. But once you as a seller have told me what the technology is, and what it can do in a sufficient level of detail that I understand its capabilities, you have effectively transferred the technology to me without any compensation!" So suppliers must consciously limit the information they provide, and as a result customers must make evaluations on highly incomplete information.

There are still other problems that arise in searching for and then evaluating external technologies, such as the problem of contamination. If the customer is a very large company, and the supplier is a very small company, this David-and-Goliath situation may make a jury very sympathetic to the small company, even if the large company developed its approach in a completely independent manner. Small companies have to worry as well. Some of their best ideas and technology may not be well protected. An in-depth discussion with a large company working in a related area may allow the large firm to understand and imitate much of the small company's value without directly infringing on its protected IP.

The desire to avoid contamination causes both large and small companies to adopt numerous practices to minimize the risk. These practices also, however, reduce the ability to leverage Open Innovation. Yet, in a world of Open Innovation, there is too much good stuff available on the outside to simply ignore. How can companies identify potentially valuable external ideas, and how can they access those ideas without compromising their own internal development activities within that general area?

Other challenges include bringing in new, nonobvious sources of information. If one simply rounds up the usual suspects, the chance

ISSUES IN ACCESSING
EXTERNAL INFORMATION

- Managing and protecting identity

- Managing contamination risk

- Identifying useful, nonobvious sources

- Fostering a two-sided market

- Scaling efficiently with volume

of learning new ideas is more limited. And one must also develop a plentiful source of ideas and join them to a plentiful group of potential buyers to facilitate a two-sided market. Finally, one must be able to scale up the operation to conduct business efficiently as volume grows. The box entitled "Issues in Accessing External Information" summarizes these challenges.

CHALLENGES IN ENABLING OTHERS TO
USE INTERNALLY GENERATED IDEAS

Broadening and deepening the search for external technologies is only half of a truly Open Innovation process. Opening up the other half of the process requires a company to work with others to take internal ideas out of itself for use in those other businesses. Some of the problems are the same as those noted for searching for external technologies, only now the perspective from which the company views the transaction has shifted, from buyer to seller. However, some additional problems also arise.

One of the most interesting problems is the problem of success. If an internal idea or technology is taken outside and turns out to be very valuable, that success raises a number of issues. A natural reaction in hindsight is to say, "Why didn't we keep that great idea inside?"

Instead of appreciating whatever portion the originating company received of the external success (through royalties or equity or other compensation), the second-guessers implicitly or explicitly believe that the originating company should have had 100 percent of the success.

This hindsight view is wrong-headed, because it ignores the biased and myopic nature of any successful business model.[1] If the company's business model could have leveraged the idea or technology within the existing business model, then the naysayers would have a point. Far more often, though, the external success of an internally unutilized idea is due to the substantial value added by the receiving party—not least of which is a very different business model used to commercialize the idea. When Xerox licensed its Ethernet technology to Robert Metcalfe for $1,000 in 1979, it was already using that technology inside its own copiers and printers. But to turn Ethernet into an industry standard and then develop all of the necessary products to connect IBM PCs with HP LaserJet printers, not to mention Unix workstations with laser printers, was very far from anything that Xerox was interested in or capable of doing. By creating 3Com and its business model, Metcalfe and others added tremendous value to the initial technology.

The naysayer attitude has another terrible effect: it is deeply anti-innovative. In this view of the world, it is far better to bury a potentially valuable technology than it is to let someone else utilize it and share the profits with you. This is socially very inefficient. It also denies the inventors and developers of the idea the chance to see their work in use in the wider world and to learn from the users' experience with that idea. It also eliminates the company's ability to learn from what other companies did to create value from that idea, which might suggest a direction for the company's own business and business model to evolve. Lastly, burying a technology is unlikely to work. Talented people enjoy many options outside a firm that is stifling their potential. At some point, they may simply leave and go elsewhere.

So, for both dimensions of Open Innovation, the outside-in dimension and the inside-out dimension, we need new processes to manage challenges. One solution that has arisen in response to these issues is the growing array of innovation intermediaries.

INTERMEDIARIES: ONE SOLUTION TO
THE ARROW INFORMATION PARADOX

Fortunately, a number of recently organized companies have focused their own business on helping companies implement various facets of Open Innovation. I think of them as *innovation intermediaries*, because their function either helps innovators use external ideas more rapidly or helps inventors find more markets where their own ideas can be used by others to mutual benefit. The presence of these firms enables other companies to explore the market for ideas without getting in over their heads, since the intermediaries can act as guides to help those other companies along the trail.

Being an innovation intermediary is not an easy business. There are many challenges that companies must face if they are to work effectively as intermediaries. One issue every intermediary must face is how to help its clients define the problem that needs to be solved. This definition must be sufficiently clear to outsiders that they can recognize whether they know enough to answer the problem, without being so clear as to reveal sensitive client information (a variation of Arrow's Information Paradox discussed earlier). A second issue that every intermediary must manage is the problem of identity: whether and when to disclose the identity of one party to the other party. Companies might prefer to remain anonymous for as long as possible, yet in some circumstances a buyer or seller might be unwilling to complete the transaction unless it knows who the other party is. A third issue for intermediaries is how to demonstrate the value of their service to their clients. Other processes, beyond the control of the intermediary, must occur in order for an idea or technology to become valuable, so how can one measure the contribution of the intermediary to whatever value was subsequently created? A fourth issue for innovation intermediaries is how to create or access a two-sided market, with lots of buyers and lots of sellers. When markets are thick, with many sellers and many buyers, they function very well. But when markets have few buyers and sellers and are highly illiquid, they do not function nearly so well. A related fifth issue is how to establish a strong, positive reputation early on in the company's operation. Since the concept of an innovation intermediary is itself something of a novelty, how can intermediaries develop the trust and reputation necessary to convince buyers and sellers to confide in them?

There are already a number of companies that have undertaken the task of addressing these concerns. In my work on Open Innovation, I have come to know a number of these intermediaries. An overview of how these entities operate follows, with examples of a select few that illustrate the range of functions that intermediaries provide. I have chosen InnoCentive, NineSigma, Big Idea Group, the InnovationXchange, Shanghai Silicon Intellectual Property Exchange, and Ocean Tomo as six exemplary innovation intermediaries. Note that I am not claiming that each of these is destined to be another financial success like Google. These are all small companies as of this writing. It is the case, however, that each company has solved a real problem for its customers and has demonstrated the ability to help its customers open up their own innovation processes.

DIFFERENT KINDS OF INTERMEDIARIES

Before we delve into these case studies, it's important to note that intermediaries come in different forms. Some function as agents, representing one side of a transaction (here, an exchange of IP and/or technology, for instance). These agents owe their allegiance to their clients but must also develop extensive market knowledge in their chosen area of expertise, in order to be able to advise the client on how best to approach the market. Often, the agent must negotiate with the client in advance of taking an idea or need to market, in order to set the client's expectations in a realistic range of outcomes. Good agents also need to cultivate a reputation for honesty and fair dealing, lest counterparties refuse to work with them.

Other intermediates function as brokers or market makers, which try to bring parties together to achieve a transaction. In contrast to agents, they may help shape the terms of a transaction and, in some cases, even take a position in the transaction to help bring it about. Such intermediaries function like investment bankers or exchanges. They too need to cultivate extensive market knowledge and a positive reputation. But their allegiance is not only to the client; they also care about the exchange environment in which the transaction occurs.

Table 6-1 arrays the selected intermediaries according to the primary function they provide.

InnoCentive

InnoCentive was a spin-off company that originated in drug maker Eli Lilly's internal R&D organization. A group within Lilly felt that the company needed to accelerate its presence on the Internet, and the associated business processes that could be enabled through intelligent use of that medium. Lilly soon launched e.Lilly, under the direction of Alpheus Bingham. Bingham, who has a PhD in chemistry, had previously been in charge of Lilly's research lab in Belgium.

One of e.Lilly's first investments was in InnoCentive. Under the early name of BountyChem, a small group within Lilly R&D thought that Lilly ought to look outside its own four walls for at least some of the innovation solutions that would fuel its product pipeline after Prozac.[2] Indeed, Lilly had accumulated a number of projects that it wanted to work on, but its internal R&D organization never seemed to get to those problems. BountyChem was a way for Lilly to elicit external solutions to these problems and get the issues resolved sooner and at a lower cost.

On further reflection, e.Lilly realized that BountyChem would attract a larger pool of problem solvers if other companies' problems

TABLE 6-1

Innovation intermediaries and their functions

Intermediary	Focus	Primary function
InnoCentive	Online exchange portal	Marketplace for technology transfer/agent
NineSigma	E-mail RFPs	Agent
Big Idea Group	Concept developer	Agent/codeveloper
InnovationXchange	Membership-based innovation community	Broker
Shanghai Silicon IP Exchange	Repository for legally obtained semiconductor IP	Broker
Ocean Tomo	IP merchant banker	Market maker

also were included. This could create an opportunity to initiate a virtuous circle, where more problems to be solved would attract more potential problem solvers. In turn, the availability of a large population of problem solvers would encourage companies to post unsolved problems with BountyChem. This virtuous circle is another way to think of the two-sided market problem discussed earlier. This insight caused the project to be spun off into a standalone company that would operate independently of Lilly. Darren Carroll, a Lilly executive who had led the legal work on the successful drug Prozac, was recruited to lead the new company, InnoCentive.

InnoCentive initially launched operations in June 2001, armed with a number of chemistry problems (called "challenges" in InnoCentive's model) that came to them from Lilly. These problems were focused on different aspects of chemistry, including organic chemistry, analytic chemistry, and formulation processes for chemicals. An important distinction in these problems was that some of them required conceptual thinking exclusively (paper challenges), while others required a demonstration of a solution (so-called wet challenges). The awards for paper challenges typically ran in the range of $5,000 to $10,000, while wet challenge awards ran from $25,000 to $50,000 or more.

InnoCentive realized from the outset that many of its staff would need a deep knowledge of chemistry to formulate these challenges in ways that would encourage external people to volunteer solutions to it. For starters, companies like Lilly that were looking for solutions did *not* want any of its proprietary information to be broadcast, since it was likely that some of their competitors would access the Inno-Centive Web site at least occasionally. It took substantial knowledge (as well as some trial and error) on the part of InnoCentive's staff to overcome these obstacles and to create abstracts that were informative on the one hand, and yet did not disclose company confidential information on the other. Another concern was that Lilly staff initially had trouble defining where the external solution ended (and the award payment would then be made) and where the internal use of that solution took over. Lilly had to advise and coach clients on what constituted a "solution" warranting payment.

InnoCentive has engineered its process to provide protection against contamination to its clients seeking solutions, whom Inno-

Centive calls "seekers." Clients sign a legal agreement between themselves and InnoCentive, authorizing InnoCentive to seek solutions to certain specified challenges. InnoCentive acts as an agent for the seekers, insulating client staff from inadvertent exposure to external ideas, unless and until those ideas become paid solutions.

Once this agreement was in place, InnoCentive would provide training to the R&D chemists within the seeker company, explaining how to formulate solvable problems and how to focus a problem sufficiently to enable the InnoCentive process to work. "You can't simply say, 'Find me a cure for cancer,'" noted Jill Panetta, InnoCentive's VP of R&D.[3] Here's an example of an effectively formulated problem: Eli Lilly needed an intermediate chemical to use as a precursor chemical in a process to be implemented by the drug maker. The ability to specify the chemical composition of the successful solution allowed external solvers to realize whether they had solved the problem. The solver did not have to know Lilly's process that would use the chemical. This also prevented competitors from knowing what Lilly intended to do with the solution.

When the challenges are vetted and formulated, they are then posted on the InnoCentive Web site (www.innocentive.com). The abstract for the challenge is available to anyone viewing the site. If a person thinks that he or she can solve the problem, that person must register with InnoCentive as a solver. This registration process provides contact and identification information, which allows InnoCentive to later ensure payment of any awards to the solver. If the solver wishes to offer a solution, he or she first must sign a solver agreement, which is another legal agreement that affirms that the solver is the rightful owner of the solution, and that the solver will keep confidential any information that the seeker chooses to share.

Once a solver has executed this agreement, a private room is created on the Web site for the solver to interact with InnoCentive's staff. In this private room, the solver finds additional information about the challenge, information that the seeker does not wish to have known publicly. This information might include attributes of a successful solution. The solver's proposed solution is reviewed privately by InnoCentive's scientific staff, who work with the solver to refine the proposed solution. When a proposed solution meets the criteria specified by the seeker, it is entered as a submission. The InnoCentive staff selects the best submission, and an award is paid to the solver who made that submission. This addresses the first issue noted above, the Arrow Information paradox.

A second issue that every intermediary must manage is the problem of identity. In InnoCentive's case, the identities are revealed on both sides only after a verified submission has been accepted and an award has been paid. InnoCentive verifies the identity of the solver and receives a further agreement indemnifying the seeker and InnoCentive against any false information provided by the solver. InnoCentive then makes payment of the award to the solver and invoices the seeker for that amount, plus InnoCentive's own success fee.

A third issue for intermediaries is how to demonstrate the value of their service to their clients. InnoCentive had the good fortune of a close relationship with its parent company, Lilly, after its spin-off, which allowed InnoCentive to gain some insight into the benefit provided by its services to a seeker client. Twelve of Lilly's challenges that were posted on the InnoCentive Web site in June 2001 received solutions within the subsequent year. InnoCentive received eighty-two submissions for these problems from solvers in sixteen countries and paid awards totaling $333,500 for them.

InnoCentive worked with Lilly to estimate the total value delivered by these solutions. Lilly's total costs for obtaining answers to these challenges, including awards to solvers and payments to InnoCentive, amounted to $430,000. The value of these solutions to Lilly was estimated to be $8.8 million, with $600,000 of that being the cost of doing the equivalent work internally, and $8.2 million being the incremental revenue that these solutions created for Lilly. This yielded a cost-benefit ratio of 20:1.[4]

While this evidence suggests an impressive return on investment, InnoCentive's experience with its seeker clients has been nonetheless quite uneven. For reasons that are not well understood, InnoCentive's first calendar quarter in 2002 was particularly low and was low again in 2003. The uneven repeat business from these clients, even from Lilly, was hard to explain.[5] In 2005, however, InnoCentive received an equity investment from a venture capital firm, giving it a second strategic investor (in addition to Lilly), as well as more capital to support its business growth.

A fourth issue for innovation intermediaries is how to create or access a two-sided market. While InnoCentive had a number of seekers early on, it would not have much of value to offer those seekers unless it could create a large, growing community of solvers. As of September 2003, InnoCentive had more than twenty-five thousand independent individuals who had registered as solvers on its Web site. Of these twenty-five thousand, more than seven thousand had signed agreements with InnoCentive, and more than fifteen thousand private rooms had been opened (some solvers worked on more than one challenge). Over seven hundred fifty submissions had been received, and twenty-five awards had been paid out. By the summer 2005, the solver community had grown to over eighty thousand.[6]

In a Web-mediated model like that of InnoCentive, the number of solvers is one important attribute for the company to manage. A second attribute is the diversity of solvers, since a more diverse community is likely to provide a wider range of possible solutions. Thus, it need not be the case that solvers outside a client like Lilly are "smarter" than the people within Lilly; it may simply be that this more diverse community collectively sees a wider range of ways to address nagging unsolved problems that clients like Lilly have. In InnoCentive's case, these solvers came from a diverse range of work experience and life stages, including the following:

Contract laboratories

Retirees

Students

University faculty

Small pharmaceutical firms

Biotech firms

Research institutes

Other industries

Other scientific disciplines

The solvers were also geographically diverse, coming from six of the seven continents and representing more than forty countries.

Table 6-2 shows the percentage of solvers from each country or region, in the twelve months ending in June 2002 and June 2003.

As shown in the table, a large portion of the solver community was outside the United States and western Europe, in places that Lilly had historically ignored for its own R&D, such as India, China, and Russia. In fact, those three countries were second, third, and fourth, respectively, after the United States in the number of solvers registered with InnoCentive, as of July 1, 2003. These people receive different scientific training than Lilly's staff and have encountered very different kinds of problems in their own experience.

This diverse community of solvers did not arise by accident. InnoCentive had an active recruitment effort to build this community. Under the leadership of Ali Hussein, InnoCentive had forged relationships and signed agreements with the leading academic scientific centers in India, China, and Russia. In India, for example, there was a signed agreement with the CSIR, India's central authority for academic science. This agreement enrolled more than two thousand of India's leading academic scientists in the InnoCentive system.[7] Similarly, in China, InnoCentive had signed an agreement with the Chinese Academy of Sciences. The blessing of this prestigious academy helped launch additional agreements with ten of China's leading ac-

TABLE 6-2

InnoCentive solvers' location by region

Country/region	2002 (%)	2003 (%)
United States	62	47
Western Europe	17	14
Russia, eastern Europe	4	9
Asia-Pacific	6	13
India, South Asia	6	11
Africa, Middle East	1	3
South America, Central America	1	2
Other	3	1

Source: InnoCentive company documents.

ademic universities. In Russia, InnoCentive began with the Russian Academy of Sciences and has since signed agreements with Moscow State University, Saint-Petersburg State University, and the other top twelve academic centers in the country.

The relationships with these leading academic centers added credibility and legitimacy to InnoCentive's efforts in each country. One surprise for InnoCentive's leaders had been the warm embrace of each country's policymakers toward the InnoCentive model. "These governments are delighted with InnoCentive's model," said Hussein. "Instead of losing much of their top talent to the West, as they had in the past, we can bring the opportunity to them, without them having to leave their home country."[8]

An article in a leading Indian business newspaper also took favorable note of the new trend in global sourcing of R&D.[9] As the article describes, a Dr. Satyam was awarded $75,000 by InnoCentive for a solution to a "wet" problem. Unlike earlier generations of scientists, who had to emigrate to the United States to apply their talents, Dr. Satyam simply logged on the Web to view the problem and then logged on again later to submit his proposed solution. He was delighted with the award. Seventy-five thousand dollars goes a long way in India.

While InnoCentive's long-term success is still to be determined, it demonstrates some important aspects of what an innovation intermediary must do to be effective. First, it must help shape the definition of the problem to be solved. Second, it must establish a process that protects confidential and proprietary information, including the identity of one or both sides of a transaction. Third, it must develop credible evidence to document its value to the parties in the transaction, both during the transaction and afterward. And fourth, given the early stage of intermediate markets in most industries today, the intermediary must help develop both sides of the market, to create greater liquidity of transactions and elicit a wider variety of possible solutions (which in turn requires a diverse pool of knowledge providers).

But InnoCentive must also decide how best to focus its business model going forward. To date, it has tried to provide a valuable service to its seeker clients, since that is where its revenue comes from under its current business model. What if the company instead turned its focus around and built its business model around its solver community of eighty thousand? Such an adjustment would pose a difficult transition phase but might offer greater long-term value.

NineSigma

A second company, NineSigma, has established a different process for supplying innovation intermediation services. To oversimplify, the company uses an extensive database of e-mail lists and works with its clients to send out targeted e-mail requests for proposal (RFPs) to subsets of its database. More recently, the company has also expanded its Web site to solicit proposals online, as well as responses to targeted e-mails.

Unlike InnoCentive, NineSigma serves a broad range of industries in addition to the life sciences, including automotive, consumer products, packaging, and engineering materials.[10] Another difference is that NineSigma publicly identifies the company seeking a solution when it solicits a proposal from its database and its Web site. Like InnoCentive, NineSigma has crafted a number of arrangements to define the problem appropriately to attract the interest of outsiders, without giving away proprietary information. As Mehran Mehregany, the former chairman and founder of NineSigma, told me: "Helping our clients define the problem is a critical part of our process. A problem that is properly defined is half-solved!"[11]

In an article, Mehregany goes on to say:

> Open innovation does not mean outsourcing R&D, nor does it mean closing down internal R&D. It is a strategy of finding and bringing in new ideas that are complementary to existing R&D projects. Open innovation removes many of the boundaries— geographical, technological and corporate—that stand in the way of new product development and new markets. Open innovation provides access to knowledge and technologies that would take years and millions of dollars to develop in-house. The approach makes it possible to shorten product development cycles and leapfrog the competition. And it makes it possible to harness so-called "disruptive technologies" instead of being blindsided by them.[12]

While these benefits may sound compelling, they actually require substantial organizational change before they can be received. As NineSigma has worked with its clients, it has come to understand that its value to those clients depends on the internal commitment of the clients to a new approach. The article quoting Mehregany continues: "Senior managements are quick to see the benefits . . . Middle managements can be a problem if they're not briefed properly on corporate objectives. For example, they may view NineSigma searches as evidence that they weren't doing their job before. They need to appreciate that open innovation makes internal R&D more strategically valuable to the company."[13]

Another virtue of NineSigma's approach is that the company is able to refine its extensive database of contacts from every search that it does. When contacts leave or change e-mail addresses, Nine-Sigma is able to update its database accordingly. When responses come in, the company is able to analyze what factors increased the likelihood of receiving a response, and that learning helps it target the outreach effort more precisely for the next project. What seems like a commodity function (the ability to send a mass e-mail) becomes a highly targeted, highly effective capability (the ability to match searches for ideas to lists that have been carefully refined, due to prior search experiences).

NineSigma has also received an equity investment from Procter & Gamble, giving it additional respectability and further capital to grow its business.

Big Idea Group

The InnoCentive and NineSigma cases might suggest that innovation intermediaries largely are confined to high-technology industries like chemistry and the life sciences (though NineSigma also serves other industries). Big Idea Group (BIG), however, is an innovation intermediary that is avowedly low-tech in its business. It acts as an intermediary in the toy industry and in the home-and-garden industry.[14]

Michael Collins founded BIG in July 2000. A former toy-industry inventor himself, Collins understood both toys and the frustrations of inventors who had a very hard time breaking into that industry. He founded BIG on the insight that in industries like toys and home-and-garden equipment, there were few, if any, economies of scale in innovation: "Collins believed that other [industries] had virtually flat product development scale curves, implying no real advantage to spending $20 million versus $2,000 to develop a prototype."[15]

In these industries with few economies of scale, there were a huge number of potential inventions in the minds of individual inventors and small companies. But this wealth of potential supply had no clear pathway to the market. And large toy companies and home-and-garden companies typically disdained the approaches of would-be inventors, as did the large retailers that sold these products. In fact, Collins negotiated a relationship with Toys"R"Us (TRU) to receive all of the submissions that individuals and small companies made to it, precisely because TRU had no process to deal with those submissions itself. As Andy Gatto, senior VP of TRU, stated, "We are inundated with requests from consumers and amateur inventors who think they have great ideas . . . If I were to entertain every one of these propositions that came in by mail or fax, I would need a staff of people just to handle this segment . . . I would find it difficult to give this high enough priority to do it internally."[16]

Collins created a number of processes to seek out prospective inventors and their ideas. One process is the BIG Idea Hunts, in which BIG invites inventors to local hotels or conference spaces to demonstrate their ideas to a panel of judges assembled by BIG. BIG also publishes a periodic newsletter and e-mail message to reach out to inventors. Its Web site provides another means to solicit ideas from inventors. Through these mechanisms, BIG now receives hundreds of ideas every month.

These ideas, however, arrive in many conditions, most of which are very far from being ready for market. Not only does BIG screen these submitted ideas, it works to develop and polish those that have potential merit by using its knowledge of the toy industry or the home-and-garden industry. These improvements truly add value to the concept and often include competitive research, repositioning the product, design and engineering enhancements, and even finding lower-cost vendors for parts of the idea. BIG often takes three to

six months to refine the ideas that have real potential before present-
ing them to prospective companies and retailers.

BIG helps inventors obtain stronger protection for their ideas by
generating trademarks, copyrights, and other protections. BIG splits
up-front advances and ongoing royalties with inventors in an open
and transparent way so that the inventor doesn't have to worry about
surprises or "creative accounting." While the inventor may only end
up with 40 percent to 50 percent of the value from his or her invention, that is a smaller piece of a much larger pie, due to the many
value-adds provided by BIG.

While the toy industry was BIG's first industry, it has more recently expanded into the home-and-garden industry as well. The
BIG model has potential to work in a variety of industries where
there are few economies of scale in invention, where the company
can build close relations with some of the key players in the industry,
and where it can leverage the creativity of external inventors. BIG
demonstrates the power of being able to connect inventors and their
ideas with markets and their requirements. BIG is much more than a
switchboard. It actively develops promising ideas further so that
projects are presented in ways that conform to the needs of potential
customers. As BIG expands its network of inventors, and as it deep-
ens its relationships with its customers, it will develop an even
stronger position in its market.

The InnovationXchange

The previous intermediaries in this chapter all start with a known
problem that the client wants to solve, or a specific solution that is
looking for a problem and a market. The InnovationXchange (IXC) is
a different kind of intermediary, in that it helps member companies
share poorly defined needs and tries to match them to technologies
and initiatives residing in other member companies. The IXC is
based in Australia and offered by the Australian Industry Group, one
of the country's oldest industry associations. It is the brainchild of
John Wolpert, who previously managed IBM's alphaWorks team and
Extreme Blue innovation lab. He believed that innovation processes

themselves had to be innovated anew. Inspired by academic research in Open Innovation, and having experienced firsthand some of the problems of Open Innovation, Wolpert set out to create a new kind of organization. This organization would help companies identify and successfully use Open Innovation in their businesses.

Figure 6-1 provides a schematic depiction of the Innovation-Xchange model.

The IXC model constructs an innovation network among its member companies. In the first year of its operation, there were eleven member companies, all of which were operating in Australia and the United States. Each member company is assigned a trusted intermediary (TI). The TI is an employee of the IXC, but his or her role is to work as a part of the member firm's inside business development, research, or commercialization team. Ideally, the TI becomes a "fly on the wall," listening intently to uncover the problems and needs of the member organization.

TIs sign agreements with the IXC that prevent them from owning or holding any IP rights in any of the work that they do, making them

FIGURE 6-1

The InnovationXchange model

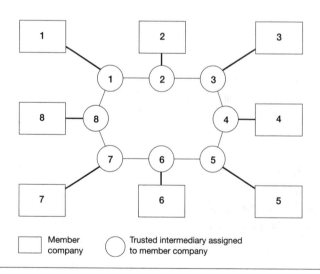

Member company   Trusted intermediary assigned to member company

true intermediaries. The TIs also sign strict confidentiality agreements to protect the knowledge of the member companies with which they work. These legal protections are joined by social investments, as the TIs spend significant time with the member company, building greater understanding and trust. The IXC describes the trust-building process as "the three Bs": beer, bonding, and being there. One to three months' time is typically required to achieve sufficient trust to become truly engaged. The role of the TI is to remain on the member's team as day-to-day commercialization support staff indefinitely.

The final piece of the model is that the TIs meet and exchange ideas with each other frequently. After gaining acceptance at the member companies and learning about needs and opportunities, the TIs then search their secure knowledge base of member information and meet with their fellow TIs to look for possible matches between the member companies. Because the TIs have forfeited any rights to the members' IP, and because of the TIs' confidentiality agreements, they can share openly with one another, without worrying about proprietary information leaking outside the network. This fosters a much richer discussion. The information being shared by TIs is not even disclosed to the member organizations except through a stepwise disclosure process called an "opportunity brief," and then only in cases where TIs have identified what they consider to be a mutually beneficial connection between the different parties.

An unintended result of the model is that member companies are using their TIs in ways that were not contemplated by the IXC's founders. The TIs have turned out to be useful in scouting connections not only within the membership but also in scouting potential connections with firms outside the membership. These external investigations have become part of the day-to-day work of the TIs. Today, more investigations are undertaken by intermediaries on nonmembers than on members.

Here is an example of how this works. A member business development team plans to find partners for a specific new initiative. The member leader asks the assigned TI to investigate companies that might make good partners, pointing out two nonmember companies that the team would like to consider. (The TI has already been in the planning meetings for this initiative and understands the internal intentions and dynamics around it.) The TI first searches the IXC confidential knowledge base of information from the other members. He contacts the TIs of those members that he finds might be of interest. He also brings up the initiative in the regularly scheduled intermediary online summit, so that the TI network is aware of the initiative.

But the TI also conducts a nonmember search, and in addition to the two companies the leader asked him to contact, he finds two more. He contacts these targeted nonmember companies (with permission of his member leader), arranges to meet them, and signs their nondisclosure agreement. In this case, the member leader did not wish to be identified by the prospective partners before the intermediary found out whether there was a good fit, so he identified himself as an employee of the IXC and said that he was looking into the company because it was of interest to the IXC membership. The TI thus cloaks the identity of the interested member at this stage.

There is also a speed advantage, as the TI signs the nonmember NDA in short order, typically within a week. Were the member firm itself to initiate discussions at this stage, the NDA process alone might take months to complete, making the search process much slower and more cumbersome. Once under NDA with the target company, the TI was able to assess within two weeks whether the firms should be brought together. In the cases where a fit was found, the TI then followed the standard IXC process to engage the member and the target.

The results of the first year's activities of the IXC have been regarded as highly successful. In the first year of the Innovation-Xchange, there were twenty-four opportunities identified among the eleven member companies, ten of which had already progressed to the engagement stage, where negotiations were under way to complete the transaction, with one transaction already completed and generating value. Some of these opportunities moved technology from one member firm to another. Other opportunities involved the creation of a joint initiative between two member companies. Others brought in outside, nonmember firms into a particular project.

These results are surprising. With the exception of one global company, these firms all operated in Australia, and most operated in

Sydney or Melbourne. The senior leaders of the member firms knew each other already. So there was a social network already in place. How is it then that the IXC identified so many new opportunities for the member firms, when these firms and their senior people were already known to one another? The IXC model seems to be the answer. One key to the model's success is the creation of a new role, the trusted intermediary, and the TI's twin duties of working closely within the member company and collaborating with other TIs operating inside other companies. The IXC model lets members exchange tacit knowledge and confidential knowledge, filtered through the TI network. Even in the relatively small world of Australian business in Sydney and Melbourne, this model is finding new opportunities and adding new value.

Based on this initial success, the IXC has expanded its operations internationally. It doubled its membership within a few months of completing its pilot phase in June 2005, and it is recruiting more member firms in Australia, with most of the eleven firms in the initial network continuing their membership postpilot.[17] In July 2005, the company began to locate TIs within the United States as well. This will further expand the IXC's network, and that should greatly increase the number of potential areas to look for opportunities. It does, however, challenge the ability of TIs to collaborate closely with one another. The IXC is developing a series of Web-based tools to facilitate greater virtual cooperation between its TIs. It will be interesting to see whether the trust-building effects of "the three Bs" can be achieved through advanced information technology solutions, or if the need for the three Bs limits the model's ability to scale.

Shanghai Silicon Intellectual Property Exchange

When one thinks of intellectual property in China, one's first impressions turn to phony Rolex watches, $2 DVDs, and other evidence of extensive piracy. One certainly does not expect to find some leading-edge practices that promote the identification and legal exchange of intellectual property in an industry like semiconductors. Yet this is exactly the purpose of the Shanghai Silicon Intellectual Property Exchange (SSIPEX).

SSIPEX is one of three centers created in China to facilitate the legal exchange of semiconductor IP. A sister center, called ICC, focuses on providing legal access for Chinese companies to design services platforms in semiconductors, such as electronic design automation tools. Another sister organization, called ICRD, focuses on providing Chinese firms with authorized access to manufacturing process platforms, to help them build the designs they develop.

SSIPEX focuses on collecting, evaluating, and disseminating the technologies that bridge the design of a new chip and the foundry process that makes the chip. Like its sister organizations, SSIPEX is an IP intermediary. It operates by working with owners of semiconductor IP to accumulate libraries of manufacturing design tools, reference designs, and other useful knowledge. It then invites local Chinese companies to come in and try out this IP. If the Chinese company finds the IP useful, then SSIPEX helps broker a license of the IP to the Chinese company. About 70 percent of the IP at SSIPEX comes from outside China, while 30 percent comes from within the country. It currently boasts more than three thousand individual pieces of semiconductor IP, making it the second-largest commercial repository of its kind in the world. Unlike the repositories of private foundry firms (such as SMIC, the largest foundry in China), SSIPEX is open to all members, regardless of which foundry members choose to use for building their designs.

Though it may seem odd to western readers, the reason that SSIPEX and its sister organizations were created is that the Chinese government strongly supports IP. A key issue that the Chinese government worries about is that many entrepreneurs do not know much about IP and how best to use it. The government hopes that these centers will allow entrepreneurs to use IP more effectively. SSIPEX was built in 2003. It was funded by the Shanghai city government with 30 million renminbi and the national government's Ministry of Industry and Information (MII) with 10 million renminbi (this combined funding amounts to about US$5 million).

SSIPEX's revenue comes from three sources: first, a membership fee charged to companies that want access to the IP; second, a fee charged to IP owners that want to display their IP, and third, transaction fees for brokerage transactions between the members and the IP owners. While the first two sources have been the dominant sources of revenue to date, SSIPEX expects the third source to grow as more Chinese companies learn about the exchange's services and understand how to use them. The organization is only two years old as of this writing, with a budget of 10 million renminbi and a staff of thirty people.

While SSIPEX is still very young, it is beginning to make investments to add more value to its member companies, as it "kicks the tires" on the extensive library of IP at the exchange. The company now employs a handful of consultants and analysts to assist member companies. A new investment in 2006 will establish a laboratory inside SSIPEX. This laboratory will function as a "black box," such that customers can bring a sample of their design to the lab, and the lab will produce a partial layout (or other output, depending on the specific IP in question being tested). But the black box will prevent customers from seeing exactly how the output was obtained and will keep them from trying to reverse-engineer or otherwise appropriate the technology. So customers will get more detailed information on the value of the IP they are trying to use, while the IP owner will remain protected from misappropriation.

How will the SSIPEX evaluate its own performance? The answer is going to be determined by its owners: the City of Shanghai and the MII. Both want to streamline the legal transfer of IP in the semiconductor industry and make it as transparent and economical as possible. And the owners also want to promote the greater development of semiconductor IP in China, which they believe will enable the semiconductor industry to grow. In addition, SSIPEX must achieve break-even financial performance in 2006.

While the SSIPEX is an exciting experiment in innovative ways to facilitate the exchange of IP, the organization nonetheless faces some daunting challenges. One problem is that SSIPEX's customers are neither strong nor large; they are small companies. In China, many people believe that engineering labor is cheap, while capital is scarce, so they think that it is more affordable to develop IP on their own. There is no appreciation among these companies that leveraging external IP could save time and improve the quality of the resulting product. This mentality is widespread and will require extensive education before many companies will reconsider.

Another challenge is the underdeveloped legal system standing behind the legal protection of IP. SSIPEX takes careful steps to ensure that the IP it offers is legally obtained. However, it does not have the resources to monitor the usage of the IP by the small Chinese companies that are its customers. If the customers were illegally reselling or otherwise transferring the IP to others without proper authorization, SSIPEX might not know about it. And if it did detect such activity, it is unclear how effective any recourse to the Chinese courts would be. Of course, as a broker, SSIPEX might be able to avoid direct involvement and leave any legal actions to the IP owners whose rights were infringed. But if IP owners determine that SSIPEX was undermining their ownership position in China, that would damage the development of legal IP exchange in the Chinese semiconductor industry for everyone.

Ocean Tomo

In downtown Chicago, a small, growing company called Ocean Tomo is quietly laying the groundwork for a revolution in the use of patents and associated IP. Jim Malakowski, who is the chairman of Ocean Tomo, along with his colleagues at the firm, progressively are attacking the many barriers that currently hold up greater exchange of patents in the market.

Malakowski himself is no newcomer to the area of managing and valuing IP. Since he graduated from college more than twenty years ago, he has devoted his career to different aspects of how to manage these assets in a more effective way. Earlier in his career, he worked for accounting firms and advisory firms, analyzing how best to account for intangible assets like IP and know-how. These years of experience gave him firsthand knowledge of what some of the key problems were. For example, there is general consensus that patents can be valued on cost basis (how much it cost to develop the idea), an income basis (the stream of payments a firm could expect to receive from licensing), or a market basis (what similar patents have sold for in the past). Of these, the market basis is regarded as the best approach. However, patent sales are rarely published, making it nearly impossible to identify comparable sales. And there may be other terms of the sale that are idiosyncratic to that situation, which influenced the negotiations over the sale and are no longer identifiable. How can one value a patent when price information is not available?

Ocean Tomo knows the problems in managing and valuing patents, and it thinks it has come up with some of the answers. It is no exaggeration to say that Ocean Tomo is the first full-service merchant bank that is exclusively focused on transacting IP. Here are some of the services that Ocean Tomo provides to its clients (and more are on the way):

- IP merger and acquisition advice

- A patent ratings system

- Advisory services for securitizing patents and trademarks

- IP risk management products

- The Ocean Tomo 300, a publicly traded equity index based on IP assets

- The world's first live patent auction, hosted on eBay, held in San Francisco on April 5 and 6, 2006

While the array of activities is impressive, the thinking behind them is even more interesting. Ocean Tomo sees the economy shifting from physical assets to knowledge assets and understands that IP is a core element underpinning the value of knowledge assets. Thinking of patents, trademarks, and other IP as just another asset class suggests that financial instruments can be devised to allow investors to participate in this asset class. As Keith Cardoza of Ocean Tomo put it, "What if investors could invest directly in Boeing's innovations, as opposed to investing in the company's stock overall? In the near future, pension funds will treat IP as an investment class. It will be the source of a new alpha and a way to diversify one's portfolios more broadly."[18]

While most of the work that Ocean Tomo does relates to U.S. patents, many of its clients come from outside the United States. Foreign companies have noted the increasing strength of U.S. patents, which we first examined in chapter 3. They increasingly seek advice on whether they ought to purchase U.S. patent portfolios from companies that might be willing to sell. Ocean Tomo helps these companies identify which patent portfolios might be of interest and what an appropriate price might be to offer for the portfolio.

While the company has offered advice in this area for years, more recently it has decided to put its money on the line as well. In 2005, the company started soliciting investments for a $200 million fund that would provide financing (loans, acquisitions, recapitalizations, buyouts) to companies that, according to Ocean Tomo's methodologies, had substantially undervalued IP in their portfolio. This might be a firm with $100 million in revenue that is cash flow positive but is looking for capital to expand its business. Ocean Tomo would work with other lenders to obtain the IP rights as collateral for its portion of a lending facility or other investment participation to the firm. Ocean Tomo, in turn, might lend up to 25 percent of what it estimates to be the value of the company's IP.[19]

One of the first investors the company solicited for investing in the fund was Ross Perot. To the company's great delight, Perot elected to provide the entire $200 million for the fund, which is now up and running. This level of investment from a respected investor confers significant credibility on Ocean Tomo's concept. If returns are as good as expected, it is likely that others will soon emulate the model, marking an important step toward a more liquid market for securitizing IP as collateral for financial transactions.

Another idea that has not yet gotten off the ground is the sale/license back vehicle for intellectual property. As we noted in earlier chapters, most companies don't fully use the patents that they own. Yet there are great inefficiencies that limit a company's ability to get more value from its unutilized patents and other IP.

The Ocean Tomo sale/license back concept could change all that. The company would sell its unused IP to Ocean Tomo, which pays cash for the IP. Ocean Tomo also licenses back the IP to the company on a nonexclusive basis, so that the company can continue to operate its business as before.

How would Ocean Tomo benefit from this? As Malakowski points out, "Patents don't just reduce risk of infringement when aggregated. They also increase in value. It's the only asset class that does that."[20] Why? Because it is often easy to invent around a single patent, by doing something just different enough to avoid infringement. When one has many related patents, the chance of inventing around all of them is much lower. This makes the protection provided much stronger and increases the economic value of the group of patents.

By aggregating IP across hundreds of transactions like this, Ocean Tomo now has a broad portfolio of more valuable IP to offer to companies for many different kinds of transactions. One such offer might be in the form of IP insurance, to provide greater protection against infringement claims of third parties.

In a world of stronger patents, where patent trolls are cropping up in unexpected places and threatening businesses, it might be good to get to know the folks at Ocean Tomo.

THE VALUE OF INNOVATION INTERMEDIARIES

Searching for external technologies to use in your business is a complex task that requires new organizational processes in order to succeed. While there is no substitute for effective internal processes to do this, this chapter has documented a variety of innovation intermediaries that can help companies search outside. In this way, companies can test the idea of an external search before committing significant resources to redesign internal processes to sustain searches on a consistent basis.

Table 6-3 reviews these intermediaries and considers how each addresses the challenges of accessing useful external information identified in "Issues in Accessing External Information" at the beginning of this chapter.

Each intermediary varies in its approach to managing the issues of identity, contamination, and sources of ideas and technology. Each also faces unique challenges in attracting more buyers and sellers into a thick, two-sided market where exchanges are frequent. And each faces issues in scaling its operations to handle more volume. This variety is typical in an emerging phase of a new kind of business process. It is certainly too early to speak of "best practices," as each organization is experimenting with how best to serve this new market area.

TABLE 6-3

Challenges of accessing useful external information

Intermediary	Identity	Contamination	Useful, nonobvious sources	Two-sided market	Ability to scale
InnoCentive	Buffers via public and private Internet rooms	Seeker sees only valid solutions	More than 80,000 solvers in many countries	Constrained by number of seekers	Strong on solver side; constrained on seeker side
NineSigma	Buffers via e-mail lists	Clients see only qualified responses	Numerous and diverse e-mail lists	Constrained by number of clients	High personal involvement to define problems and handle responses
Big Idea Group	Acts as agent for inventor	Bypasses internal R&D	Individual inventors; Idea Hunts	Constrained by market focus	Limited by number of "Michael Collins" on staff
Innovation-Xchange	Buffers through trusted intermediaries	Members see only valid matches	Unused ideas and intentions unlocked for new uses	Robust between members, sparse outside	Constrained by TIs ability to connect deeply with each other
SSIPEX	Serves as IP broker	No client access to source code until license	Western semiconductor IP, many small Chinese firms	Must educate Chinese companies on benefits of IP for fast time to market	Promising, if education efforts succeed
Ocean Tomo	Serves as IP merchant bank	Usual merchant bank protections; Chinese walls between functions	Latent, unutilized, and underutilized IP	Hard-to-value IP, no good pricing data; education required	Emergent phase; new IP index to help educate; has significant capital base

Yet, if you are not yet persuaded of the value of these services, consider how you would feel if your leading competitors were to start making use of them. It would be hard for you to track what they were doing early on. It is likely that they would find opportunities in new places that no one in the industry had looked at before, including you. And some of these technologies may already be quite far along in their development, so that they might reach the market in a surprisingly short period of time. Would you want to compete against that? Wouldn't you prefer to give your competitors the headache of competing against that?

Once external ideas and opportunities are located, though, more work needs to be done. Your business model needs to open up to make room for these new opportunities. In the next chapter, we will discuss some examples of companies that are crafting business models that thrive in the Open Innovation context. They may inspire your own thinking, as you search for a business model that can exploit the wealth of useful external knowledge that surrounds you.

7

IP-Enabled Business Models

The previous chapter in this book discussed the emergence of innovation intermediaries. These firms can help companies of many different sizes participate in the emerging secondary markets for innovation and IP, and craft more open business models. In this chapter, we examine a number of firms that have gone much further in participating in these secondary markets. Each of the firms profiled here has built its business model around these secondary markets. They are not agents or enabling firms in the secondary markets; they are owners that are driving the development of these markets. While there are many differences between each of the firms that follow, what connects them is the use of IP as a pure play, where the chief capability of the organization is its ability to create, own, market, and sell IP to other firms.

In reviewing these companies, we will examine how the business model itself changes when it is based on IP. For while some aspects of the business model concept apply to IP-based situations in a straightforward manner, other aspects need to be adjusted in order to apply. In this sense, IP-based business model companies are innovating on the concept of a business model in addition to their approaches to their respective businesses.

We will review three IP-based business models—Qualcomm, UTEK, and Intellectual Ventures—and see how they extend the concept of a business model at the conclusion of the chapter.

QUALCOMM

Qualcomm is a company based in Southern California that specializes in cellular and wireless telephony. It is best known for its code division multiple access (CDMA) technology, which provided a more efficient way to use scarce airwave capacity to support a higher number of cellular phone calls. Dave Mock's book *The Qualcomm Equation* recounts the impressive business model that the company has created around its key technologies and the associated IP.[1] Because Qualcomm sells its technology only through licenses (and also receives revenues from selling some custom semiconductor chips that employ its technology), it is able to generate tremendous income without deploying armies of people or billions of dollars of physical assets. In the fiscal year 2004, for example, the company generated $4.9 billion in revenue and $1.7 billion in net income with a workforce of eight thousand people (yielding a revenue per employee figure of $612,000). This is a very leveraged model![2] Also impressive is that Qualcomm's performance is "asset light," since the company creates IP and designs chips but leaves it to others to manufacture the chips. In fiscal year 2004, the company had $10.8 billion in assets, but fully $7.6 billion of those assets were in cash or cash equivalents. Subtracting the cash from the assets needed to support the business, Qualcomm earned a 53 percent return on assets for 2004.

These are impressive figures for any business and have rightly drawn attention to Qualcomm as having a new kind of business model. As I will discuss later, though, Qualcomm's model took a lot of effort to create. Its realization required a substantial amount of risk, and some luck. Its history should be kept in mind by other companies that seek to emulate the financial leverage of Qualcomm's model.

As Mock's book shows, Qualcomm did not start out to change the world of telephony. It began as a developer called Linkabit, which made a variety of specialty technologies for the government, and particularly for the military. Like many such developers, Linkabit was living from contract to contract, trying to keep its people and its overhead expenses covered and, frankly, not making much money. It was acquired by another company, and disagreements after the acquisition caused Qualcomm to spin off from the acquiring firm in July 1985. That same year, Qualcomm signed a development agreement with Omninet Corporation to help design and implement a satellite messaging system. Qualcomm was charged, in particular, with building the modem on the ground end of the system.

To Qualcomm's credit, the company thought big and went well beyond the requirements of the specific contract. It chose to use a clever technique to get more capacity out of a slice of the airwaves (called "spectrum") used to communicate with the satellite. This technology became known as code division multiple access (CDMA) technology. It was incompatible with earlier wireless telephony technologies, including one that was beginning to gain real volume in the cellular market, known as time division multiple access (TDMA). If TDMA became the dominant design, Qualcomm's technology might lose the race altogether. However, Qualcomm lacked the resources, experience, and expertise to get into the cellular phone business on its own. And none of the companies in the cellular business that Qualcomm talked to had much interest in CDMA. Those companies felt that the technology was unproven. While it looked good in theory, it might not work in practice. And there was the alternative TDMA technology that was already working.

After years of battling this frustration, Qualcomm belatedly and reluctantly got itself into the cellular phone business, for the sole purpose of proving to the world that its technology could work in practice. This meant that Qualcomm had to build the cellular phones itself. Qualcomm had to build the base stations that relayed the cellular signal from tower to tower to handle calls. Qualcomm had to finance the development and deployment of multiple demonstrations of the technology, both in Los Angeles (where Pacific Bell underwrote most of the costs of the trial in 1989) and in New York City (where NYNEX underwrote most of the costs in 1990).

In 1991, Qualcomm got its first customer, which turned out to be the Korean Electronics and Telecommunications Research

Institute. By 1993, it began to receive its first real royalty revenues for CDMA technology. That was eight years after Qualcomm started on the Omninet project. In the meantime, Qualcomm had gone public, done a secondary offering, and spent hundreds of millions of dollars developing and proving its technology. Soon after, Qualcomm had to enter into application specific integrated circuit (ASIC) semiconductor chip development, because its technical requirements for its CDMA technology were sufficiently advanced and unusual that no established chip companies could make the chips that it needed. This would require hundreds of millions of dollars more in R&D.

It was only much later, once CDMA technology was deployed at dozens of telecommunications companies around the world, that Qualcomm could back out of some of its earlier investments in vertical integration. By then, CDMA had proved itself in dozens of carriers around the world, and Qualcomm was leading efforts to advance further the technology. One landmark deal occurred when Qualcomm sold its handset and base station business to Ericsson in 1999, as part of a settlement of an IP dispute between the two companies.

So the lean, mean cash machine that is Qualcomm today evolved out of a much more vertically integrated player earlier in its history that spun off less relevant divisions and divested underperforming businesses. Qualcomm's impressive earnings today are the result of hundreds of millions of dollars invested in earlier years, and some very rocky times of poor financial returns. To its credit, the company today is more focused and seems poised to continue its leadership in cellular telephony. With the rise of the Internet and the advent of data-based networks, Qualcomm is poised to serve these markets as well with its CDMA technology. But it's worth remembering the work and investment it took to craft this IP-based business model.

Qualcomm itself may have forgotten this lesson. Another recent initiative of the company was a technology called Digital Cinema. This technology would enable digital transmission and display of the latest Hollywood movies at theaters around the world. Instead of having to create master film copies of pictures, and then distribute physical copies to thousands of theaters, Digital Cinema allowed movies to be distributed directly over data networks. This eliminated the mastering costs of film, enabled faster delivery of new films, and gave theater owners more flexibility in the movies they chose to run. Qualcomm had figured out all of the technical challenges in making digital cinema work and hoped to license its technology to every major studio—a direct replication of its technology-rich, asset-light business model of CDMA.

However, Qualcomm's efforts to establish Digital Cinema have not succeeded. The company had a target market but did not have a compelling value proposition for the movie theater owners, who had to adopt the technology. The technology got off to a promising start, receiving $14 million in licensing revenue in 2002. But then problems emerged with the business model. The theater owners could not charge their moviegoers more money for viewing a digitally produced film. Qualcomm's approach was not the only feasible approach to digitizing film production and distribution. And its insistence on a licensing business model meant that someone else would have to figure out a compelling proposition to promote the technology and convince skeptical owners that Qualcomm's was the better technology. In this instance, no one else stepped up to make the investment, and Qualcomm lost the competition. The company discontinued the initiative and wrote off the business in June 2004.

By avoiding the investment needed to deploy and implement the digital cinema system, Qualcomm may have overlooked its own history and the role that its very hands-on involvement played in the success of CDMA. Put differently, Qualcomm's experience with both CDMA and digital cinema shows that there is a lot of hard work that lies behind a successful IP-based business model.

UTEK CORPORATION

UTEK works with universities to create and launch new companies to commercialize promising university-based technologies. In contrast to the intermediaries of chapter 6, though, UTEK remains an equity owner in its ventures for an extended period of time after a transaction is completed. (This is why I have placed it in this chapter instead.) UTEK's focus as an organization is clear from its mission statement on its Web site: "UTEK Corporation is dedicated to building a strong bridge between university technology and companies that can bring useful new ideas to the marketplace. We strive to inspire our partners to push the envelope of technology to introduce new developments that improve the quality of life and create lasting value."

UTEK was founded in 1997 and is based in Florida. The primary business for UTEK is crafting technology transfer deals to take projects from the university to a preselected corporation via a specially created venture company. In doing this, the company does not invest in its own R&D. Instead, it relies on the corporate partner to invest whatever funds are needed to carry the university research through to the market.

UTEK is led by a former University of South Florida professor, Clifford Gross. Gross owns 34 percent of the company personally and functions as the CEO and chairman. Gross himself has been an inventor and has experienced firsthand the many problems that universities have in trying to commercialize their research.

To illustrate the problem that UTEK was created to solve, Gross points out that in 2003, the nation's top two hundred universities reported approximately fifteen thousand five hundred new inventions. Approximately 70 percent of these inventions went unlicensed.[3] "This is a profound waste of human creativity and invention," said Gross. Gross is passionate on this subject and recently wrote a book called *The New Idea Factory*, as well as other books that explain the process of technology transfer to entrepreneurs.[4]

What is brilliant about UTEK is how it has crafted a business model that fits comfortably with the limits of both universities and start-up companies in transferring technology from one to the other. Universities, for example, have no cash to invest themselves in commercializing a technology, and they similarly have no interest or expertise in managing equity stakes in small start-up companies. Universities like royalties, because they provide an ongoing source of financial support to the activities of the university, without exposing it to any of the liabilities associated with owning illiquid equity stakes in young companies.

Small and midsize companies also face sharp constraints in technology transfer out of universities. For starters, such companies often don't know what technology there might be that would be helpful to them. For another, they also lack the cash necessary to buy a technology from the university, even if they did identify a promising technology. They also lack the time, experience, and patience needed to negotiate with university technology-licensing offices to obtain access to a promising technology. Only large companies typically can develop these skills.

This is where UTEK enters, with a cleverly adapted business model that bridges these widely separate parties. Figure 7-1 depicts this model in action.

As the figure shows, UTEK acts as an intermediary between the university and the public technology company. While UTEK's business model doesn't invest directly in technology development or in research, one area where it does invest directly is in building connections to leading academic researchers. This extensive academic network of contacts is used to stay abreast of promising technologies under way in various universities across the United States. This is the intellectual feedstock for UTEK, when it approaches small public companies that are looking for a technology differentiator. As of mid-2005, this $7 million company had a scientific advisory board of thirty-five people! So this network is an area of real investment for

FIGURE 7-1

UTEK business model

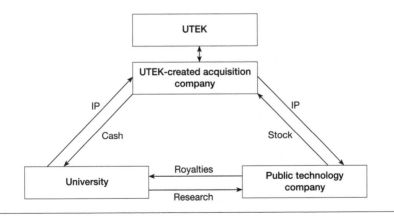

the company, and one that would be difficult for small technology companies to replicate on their own.

In parallel with its extensive academic network, UTEK identifies small technology companies that could benefit from an infusion of technology. As UTEK works with a company and identifies the company's key technical capabilities, it looks for university research projects that are close enough to the company's skills that the company could commercialize the project successfully. Since the company lacks cash to pay up front, UTEK establishes an acquisition company that takes a (typically minority) equity position in the young company upon completion of the transaction. UTEK works with the university to transfer out the technology, being careful *not* to identify the intended recipient of the technology until the closing of the transaction. UTEK's acquisition entity advances the cash to the university, receives the IP rights to the research, and then transfers those rights to the technology company recipient in return for equity. The recipient takes the commercial risks of applying a technology in its business and of creating pull through demand for the technology. Over time, the public company also may pay ongoing royalties to the university in return for rights to additional research, depending on how the agreements have been arranged.

To give a concrete example, suppose UTEK identified a young public company that's listed on the Nasdaq over-the-counter exchange in the industrial cleaning business and that's competing against big industrial cleaning companies like ARAMARK. Suppose further that, thanks to UTEK's extensive academic network, it finds a prion-based cleaning technology out of the University of Alabama. UTEK has the insight to realize that this little company would have a better investment story if it had a proprietary technology to compete against firms like ARAMARK in the cleaning industry.

So UTEK approaches the cleaning company with the opportunity to differentiate itself in the market and develop a stronger investment thesis for investors. Moreover, UTEK will take its payment in stock, not cash—say, at a rate of $10,000 a month for a year, cancelable on thirty days' notice. UTEK and the company sign an agreement.

Then UTEK approaches the University of Alabama and negotiates for exclusive rights to the prion-based cleaning technology. UTEK pays the signing fee to the university in return for the technology rights. UTEK in turn sells the rights to the technology to its client in exchange for equity in the company, per the agreement the two parties signed. Then the client company goes to its investment bank with the new technology, and the investment bank helps raise the money to pursue the technology (perhaps through a private placement or a secondary offering), and the stock value (hopefully) goes up as news of the new technology gets out. UTEK cashes out its initial position, or perhaps it keeps a portion of the equity for further appreciation. So UTEK's model requires it to function as a technology scout, a broker, and an investment banker, at various points in its business. In 2004, it had more than ten agreements with technology companies to help them identify and obtain promising university technologies for their businesses.

UTEK has been growing in the past few years by acquiring companies that were working in related areas or that were developing similar services. The latest acquisition of UTEK was in December 2004, when the company acquired EKMS. EKMS, founded in 1986, has become a leader in the field of strategic IP management under the leadership of Edward Kahn. EKMS provides technical and business expertise to help companies identify, assess, protect, and leverage IP assets to enhance market leadership and profitability.

Under Gross's leadership, UTEK is clearly on to something. Though only eight years old, UTEK generated more than $7 million in revenues and profits of more than $300,000 in 2004, supported by a staff of just twenty-nine. The financial markets see the promise of much more success, assigning the young company a market capitalization of more than $112 million (a valuation of sixteen times the 2004 revenue) as of July 26, 2005. In addition to its small staff, the company spent only $50,000 on its own R&D in 2004.[5] Much of the financial leverage we saw in the Qualcomm model is evident here as well.

INTELLECTUAL VENTURES

Intellectual Ventures was founded on the concept of "reinventing invention." Scientists Nathan Myhrvold and Edward Jung shared a vision of building a company that focused on inventing and investing in invention. While an invention company sounds unusual to modern readers, this type of venture was very popular at the beginning of the twentieth century, as evidenced by the success of Thomas Edison's laboratories in Menlo Park, New Jersey. The job category of "inventor" was used by the U.S. Census until 1946, when it was replaced with the much less compelling title of "professional worker."[6]

After many years of working with start-up companies in launching new technologies, Myhrvold, Jung, and their partners recognized that there were some fundamental problems with the invention process. For one, every technology was locked up in a single company, often a narrowly focused start-up company. This company would attempt to select and commercialize the most promising application of its technology. The focus of the venture was to execute against that application; there was little time, money, or patience for exploring alternative ways to use the technology. And the investors in the company would be sure to lock up the technology in that company, further diminishing the chances of pursuing other applications through other parties.

Since 90 percent of start-up ventures typically fail, many worthy technologies never really have a chance to realize their potential in the market. "No one else gets to try that invention," noted Intellectual Ventures' managing director Greg Gorder in an interview with me.[7]

Another set of problems arose in large companies. They viewed new technologies through the blinders of their current business and associated business model. Moreover, many companies were focusing on more near-term, and more incremental, innovation opportunities, leaving an increasingly large hole for longer-term innovations. "Sure, there are inventors in large companies, as well as in universities and in start-ups," said Gorder. "But invention is not their primary focus. We saw an opportunity to build an organization where invention was the primary focus of everything that we do."

One early initiative that Myhrvold and his colleagues undertook was the creation of ThinkFire. ThinkFire sought to help companies monetize the value of their intellectual property beyond the value they obtained by using it in their own business. "We wanted to help companies make money off of their inventions," said Gorder. "How can you make more invention happen? Find a way to help make it profitable. Right now, most companies undervalue their inventions and treat [those inventions] simply as a raw material in their own product."

To pursue this vision, Intellectual Ventures (IV) hired some of the brightest minds in business, technology, computer science, physics, biotechnology, mathematics, and intellectual property. Employees explored invention and investment opportunities in a wide variety of technology areas, ranging from software, semiconductors, electronic devices, consumer products, lasers, biotechnology, and medical devices. A particular focus of the invention process for the company was to target the areas where different scientific disciplines were coming together, such as computing and biology, and where a deep understanding of each field opened up new avenues of exploration in the other.

While the exact parameters of the IV business model are still in development, the model anticipates selling its diversified portfolio of inventions as a foundation for pursuing a variety of business models. This too was not unlike the business models of the inventors of the late nineteenth and early twentieth centuries, where devices like the automobile, lightbulb, car, and television were sold to others through business models such as licensing, spin-offs, joint ventures, selling outright, or investing to develop the inventions themselves. In general, however, the company expects to act as a technology supplier to other companies, rather than practicing the technology in its own businesses.

Even the invention model of IV had evolved in the five years since its founding. "We have changed a number of aspects of our model over the past five years," noted Gorder. "Early on, we spent more time trying to understand the exact state of the art in whatever fields we were interested in inventing in. We've learned that the state of the art is not always known, particularly from public sources. We now rely more on experts to help us here. They often know not only the public literature in their areas of expertise, but much of what is in the pipeline that has not yet reached the public."

Most companies of any size had rather bureaucratic processes for reviewing invention disclosures and deciding whether to pursue a patent filing on those disclosures. Typically, a group of research managers would meet on a monthly or quarterly basis and go over the inventions uncovered by the employees in their groups. The quality of the science in the project was a primary concern, along with whatever was known about the market and any other prior work in the area. Since patent filings were expensive, the committee would rank the inventions in priority and spend its budget on those that seemed to offer the most promise for the business. And, except in rare circumstances, companies managed their patents according to a strict budget. Once issued, most patents sat on the shelf, unused and locked away from anyone else to use.

IV has a rather different process. The invention process itself is quite distinctive. The focus of the organization around invention is unique. The people hired by the firm are unusual. The external parties play a role that differs from the one that they usually play. While the output is still a patent, it is produced through a very different process.

One key process that IV had created was the invention session. "This is the most fun part of my job," said Laurie Yoler, who was the chief development officer of the company in 2004 and 2005. "While I have to worry about a lot of other things, this is my chance to get close to the science, and to the things that we do that differentiate us from everyone else."[8] It was hard for Yoler to describe an invention session. While it was a key aspect of Intellectual Ventures' process of invention, it had to be experienced in order to be understood.

The Intellectual Ventures Invention Session

On the table cans of Diet Vanilla Coke, Pepsi One, Dr. Pepper, and bottled water sat alongside dishes of peanuts, bags of beef jerky, and plates of muffins. In display cases on one wall of the conference room, there were typewriters from the early 1900s, old cryptographic equipment, and even an assortment of firehose valves. (The aptness of the reference to the metaphor of "drinking water from a firehose," signifying the very fast rate of knowledge flows, would soon become apparent.) Around the table sat a group consisting primarily of theoretical physicists, engaged in a process created by Intellectual Ventures called an Invention Session. The lights in the conference room were dimmed because the computer projector was turned on. A variety of Web links to relevant papers and prior research were shown on the wall. Two whiteboards were in the room, both of which would shortly be filled up with notes from the group.

Nathan Myhrvold stood at one board, rapidly scribing notes on how high-energy physics could enable a new class of optical materials, using photons and plasmons to do the work formerly done by electrons. For anyone who had not studied physics in nearly thirty years, it was impossible to keep up (this was part of the "drinking from a firehose" metaphor). Even if one were current on modern physics, ideas came thick and fast. (Since the conversation moved quickly, audio recorders and periodic photos of whiteboards helped reconstruct the insights of the session later on.)

The invention session was something like a musical jam session. There were soloists, explaining their ideas to others about how properties of one material could be harnessed to provide a useful function. But no sooner was the solo done (and sometimes before the solo was done) than others in the room would riff off of the idea, weaving it into ideas of their own or questioning the basis of the concept and how it would work if deployed in a new context, far from the conventional areas where it was already accepted. And the conversation did not move in a strictly linear fashion. Inspirations would spark asides, jokes, or references to classical physicists (Hipparchus, Archimedes, Newton), early astronomers (Galileo), or even Horatio at the Bridge. Every once in a while, Myhrvold would stop the group just to capture the last set of thoughts expressed or to sum up the work done so far in the discussion.

Myhrvold is a former CTO of Microsoft, a wealthy individual, and a very successful businessman. Myhrvold is a man of many interests. He is also an intellectual polymath, with deep knowledge and interest in such disparate intellectual fields as physics, the life sciences, computer science, and photography. An inventor with more than seventeen patents to his name (as of December 2004), he was the chief instigator at the invention sessions, introducing the agenda, limiting the field of exploration (to at least some modest extent), and often pushing the group to chase ideas that were "cool as hell." Myhrvold's enthusiasm, the breadth and depth of his knowledge, and his comfort with a wide range of ideas were critical ingredients to the success of the session.

Lowell Wood also was enjoying himself. A world-class physicist from Lawrence Livermore National Laboratories, he has known Myhrvold for nearly thirty years. Most of Wood's work was done within the government, and most of that work was classified. He would sometimes allude to work done on weapons systems, high-energy lasers, electromagnetic pulses, and the like but would quickly shift back to the point he was making. He also had timely and thoughtful insights into human biology, current brain surgery techniques, and the effects of radiation on human tissue. One sensed that he found it very refreshing to be in an environment where he could explore nondefense applications of ideas, materials, and tools that he had worked with for decades.

IV typically recruits high-bandwidth experts like Wood to participate in invention sessions. Other inventors who have been publicly identified as working with Intellectual Ventures include Leroy Hood, a molecular biologist and technologist from the University of Washington; Robert Langer, a chemical and bioengineering professor at MIT; Eric Leuthardt, a neurosurgeon with St. Louis's Washington University; and Muriel Ishikawa, a physicist at Lawrence Livermore National Laboratory.[9]

Supporting Myhrvold and Wood were two other accomplished inventors. Rod Hyde and Chuck Whitmer had worked with Wood and Myhrvold, respectively, for many years. Each had made many contributions to the thinking of these two, corrected inspirations when they extended too far beyond reality, and filled in the holes left open by the rapid movement of conversation in the discussion. While Myhrvold and Wood did much of the talking, both listened with respect when Hyde or Whitmer had something to say. While the ideas were far ranging and sometimes outrageous, the prior experience of the participants and the recent research done to prepare for the session provided substantial data to anchor the participants and help them ground their insights.

Clarence (Casey) Tegreene was also tracking the output of the session. He is the chief patent counsel of Intellectual Ventures. Unlike most patent counsels, though, Tegreene is also an inventor himself, having been chief technology officer of Microvision. He held fifteen patents himself as of December 2004. "I love this guy. This is a patent attorney who also has his own invention ideas!" exuded Myhrvold.[10] Although Tegreene was a very active participant and contributor, that was just the tip of the iceberg of his involvement.

Tegreene had done a great deal of work before the session to prepare the materials for the day's discussions. He worked with Corrie Vaa, a then-recent PhD physicist from Stony Brook University (and who had just passed the patent bar exam). Together, they had conducted a thorough scan of research papers in areas of interest to the topics to be explored in the session. Tegreene would frequently reference papers that had been reviewed before the session. Vaa could pull up papers in near real time when it helped advance the discussion. And together, they had the task of capturing specific invention proposals from the session and then performing the arduous work of prioritizing, researching, and finally writing up any patents that might flow from the day's activities. "There is a real art to pulling one of these sessions together," stated Tegreene. "We limit the number of people in the room, and we are careful to control the mix of inventors to noninventing observers. If you get too many of the latter, the inventors feel like they are on stage, instead of being engaged in an invention process."

"Today's session is something of an outlier," said Tegreene. "Today, everyone in the room is a theoretical physicist. Usually, we try to mix up two or three types of experts in a single session and target the interstices between technology areas. Today is also unusual because it is a 'high-reach' session, dedicated to exploring ideas that are pretty far out there." Tegreene explained that a typical daylong invention session might generate as many as a hundred ideas ("plus or minus fifty" he added). It was not economical to pursue each invention, so he and his team would have to sift among the many inventions to select those to pursue further. The quality of the idea, the amount of material on hand to support the filing, and the market potential were all criteria evaluated in this decision.

Those selected would be compared to the prior art in the area and would also be assessed in terms of the market potential that seemed addressable with the invention. While not an exact science, these criteria enabled Tegreene to set priorities for the selected inventions and to decide which of those to pursue with patent applications.

"We try to build a portfolio of patent applications," he commented. "Some of these might be 'low reach' and be something that might turn into a real business in a short period of time. Other applications look farther out. We also try to build a range of application areas into our portfolio."

Building a Patent Capability

One Intellectual Ventures patent had already come out, a patent on a new device for converting analog signals to digital signals. "Even though this came out quickly, it is a really cool device," claimed Myhrvold. "Typically, it takes us two to three years for our patents to work their way through the process." The company sought to minimize the investment of time and resources needed to demonstrate the technology in practice. While they had to do a certain amount of this work in order to receive a patent, they relied on others to do the further development necessary to make the technology work in a product or service.[11]

The key assets in Intellectual Ventures' approach to inventing are the knowledge of the patent process, the contacts with various polymaths who have the breadth and high bandwidth to invent prolifically, and the internal staff to recognize, filter, and patent the resulting outputs. This will enable the company to avoid having to hire engineers, developers, manufacturing and operations staff, marketers, and a sales force. If and when the revenues start flowing into IV, it should scale up revenues very efficiently and generate a lot of cash. It is easy to see why Myhrvold and his colleagues are excited by this new business model. IV has reportedly raised hundreds of millions of dollars in capital from external investors (though the company will not confirm the amount of money raised), who evidently are excited about this business model as well.[12]

Internal invention sessions are not the only means by which Intellectual Ventures obtains patents. The company is also an active buyer of external patent portfolios. It participated in the Commerce One patent auction in December 2004 but declined to match the winning $15.5 million bid of Novell for the Commerce One patents. Intellectual Ventures attends bankruptcy auctions of failed start-up companies as one source of new patent portfolios. It was able to purchase all of General Magic's patents (in computer software and semiconductors) for $300,000. Intellectual Ventures is able to be such an active buyer because of the capital it has raised from its external investors.

Notwithstanding all of its innovative approaches to developing IP internally and acquiring it externally, to date the company does not have a business model through which it will commercialize its patent portfolio. In part, this lack of an articulated business model is by design: IV does not want to tip off others prematurely about how it plans to convert its IP into money. That could drive up its costs of acquiring IP (if, e.g., others began to bid against it) and might complicate its subsequent transactions.

However, one can surmise that the company has already staked out a couple of approaches to the problem of converting patents into cash. First, the company hopes to develop licensing packages for potential users of its patents. Intellectual Ventures' breadth and scope of patents might enable it to be a one-stop shop for companies looking to enter into a new area. Second, the company generally hopes to avoid having to practice any of its patents directly in its own businesses, which should strengthen its bargaining position versus potential licensees of its technologies. This stronger position in turn should help IV return capital more quickly to its investors.

However, these two approaches to the business model may contain the potential for problems, as we saw in the Qualcomm example. Qualcomm does a thriving business in licensing today. But its ability to earn those monies derives directly from the learning and know-how it accumulated from practicing the technology itself for many years. Indeed, the company's patents were thought to be worth very little, because the underlying CDMA technology was viewed as being infeasible. The company had to invest very large sums of money to develop and demonstrate the technology in order to demonstrate the technology's value. Only then were the company's patents of value to others.

Because IV does not intend to practice its technologies, it may lack some of the requisite know-how and experience to demonstrate the value of its patent portfolio in a given area to potential licensees. And if licensees cannot make the technology work, IV will lack the capability to help its licensees use its patents effectively. The high-bandwidth polymaths at IV will have to invent some answers to these business model challenges to reach their ambitious goals for the company and reward their investors.

REVISITING THE BUSINESS
MODEL CONCEPT

We now return to the concept of a business model to see how these firms are altering the concept as they pursue these pure-play IP approaches. Recall that a business model fulfills two critical functions: it *creates value* for the business offering to the value chain that the focal firm seeks to serve (the firm's suppliers, customers, distribution partners, etc.), and it enables the firm to *capture a portion of the value* created for itself. While business models are attracting more academic attention, the definition that Richard Rosenbloom and I developed is consistent with most of the academic research on the topic.[13] For ease of reference, this definition is reviewed here.

The six functions of a business model are to:

1. Articulate the **value proposition**—that is, the value created for users by the offering.

2. Identify a **market segment**—that is, the users to whom the offering and its purpose are useful.

3. Define the structure of the **value chain** required by the firm to create and distribute the offering, and determine the complementary assets needed to support the firm's position in this chain.

4. Specify the revenue generation mechanisms for the firm, and estimate the **cost structure** and **profit potential** of producing the offering, given the value proposition and value-chain structure chosen.

5. Describe the position of the firm within the **value network**, linking suppliers and customers, including identification of potential complementors and competitors.

6. Formulate the *competitive strategy* by which the innovating firm will gain and hold an advantage over rivals.

With the rise of intermediate markets for technology and innovation, based on the IP of the focal organization, one must revisit this business model concept. To be sure, IP represents a clear way to capture value from innovation. As noted in chapter 4, IP can also be managed according to a technology life cycle and assist in the value creation portion of the business model as well. But intermediate markets by their definition imply that the owner of IP is not selling a product or service in a final product market. Some other party or parties are performing that function. At least some elements of the six-attribute working definition of a business model may be harder to develop in the context of an intermediate market for innovations, where others must finish the task of commercializing the offerings.

Consider the first two attributes, the development of a value proposition and a target market for the offering. These are clear for a final product or service but are much less so for an intermediate market exchange. How can IP owners target their IP development efforts to support the business model when the ultimate use of the IP will depend on the customer that licenses or purchases the IP? Similarly, how can one establish the value proposition to the end customer when the area of use is not yet known? More generally, how can one establish the value of the IP when its use is not yet known?

Armed with a better understanding of business models from Qualcomm, UTEK, Intellectual Ventures, and (from chapter 2) open source software, we can now address this question. Qualcomm clearly knew the general use for which it wanted to develop and sell its IP: CDMA-based wireless telephony. And it had a clear value proposition to carriers: offering greater capacity of calls for the amount of spectrum the carriers possessed. However, it took eight years for Qualcomm to obtain significant royalties from this technology. To obtain those royalties, Qualcomm had to build out its entire network on a city level on at least two different occasions (the trials in Los Angeles and New York City). These trials demonstrated to potential customers that the theoretical potential of its CDMA technology translated into better capacity when deployed in the real world. Without Qualcomm's demonstrations, the many criticisms leveled at its technology by firms backing rival technologies might never have been resolved. And Qualcomm had to be opportunistic in identifying its initial customers. Its first large design win was in Korea, which was far from the first market the company considered when it began developing the technology (after all, the first demonstrations were in Los Angeles and New York). Qualcomm's own recent experience with Digital Cinema suggests that it now conceives itself to be a pure-play IP supplier. It may be that Qualcomm management has forgotten the long, hard work from its own history that made its IP-based business model viable.

Unlike Qualcomm, UTEK's business model looks at technology from a capital markets point of view, not from a technology-centric point of view. The model requires UTEK to identify potential clients whose stock price is going nowhere and would be enhanced by a technology that would differentiate the stock. UTEK scouts for such potential technologies among its surprisingly large advisory board and vast university connections. Once a promising technology is found for a client, UTEK helps the client develop a value proposition to differentiate the client's technology and enhance its stock value. UTEK takes the bulk of its fees in its client's stock, which it holds for an extended period of time, so it benefits only if and when the stock price goes up. When its client's stock price rises, UTEK is then able to exit its investment.

In UTEK's model, the real value of innovation and IP is determined by the rise in the value of the company that receives the university technology and then commercializes it. This is the polar opposite of "technology for technology's sake." And UTEK's incentives are aligned with its clients, so that it wins when its clients' stock price rises.

Intellectual Ventures has not licensed any of its inventions as of this writing, so it has not yet fully deployed its business model. Like UTEK, the company eschews much internal development of its technologies (its focus is on invention, not development), so it is leaving the job of constructing a value proposition with its technologies to its prospective licensees. For IV to be successful, its licensees will have to grapple with selecting one or more target markets and developing powerful value propositions for those markets. Then, it will have to invest to demonstrate the value to a skeptical market, which may possess alternative technologies to address those needs.

This means that IV will have to think hard about the business models of its licensees in order to craft a viable business model for itself. For example, IV will have to be able to price its inventions high enough to be attractive to generate returns for its investors but low enough to enable the business models of its licensees to be viable. Unlike UTEK, IV does not wish to take equity from its licensees; it may therefore have to structure the timing of its royalty payments so that licensees will have cash available to fund the development of the technologies.

Similarly, IV will have to think hard about how the value chain will fund any required development work needed to incorporate its inventions into commercial offerings. It is quite possible that IV's inventions will face approaches that will compete with them for a future market application. In this case, the inventions will require some development investment to be tested in competition with these alternative technologies, similar to the tests that Qualcomm's CDMA inventions had to perform, before a viable market is found. And someone will have to bear the risks of those trials and pay for that development investment.

This is not an idle concern for IV. IV will only get significant royalties if its technologies prevail in these competitions. Its ability to remain a pure-play invention supplier may be challenged by the need for some party in the overall IV value chain to pay for the development work necessary to demonstrate the value of IV's inventions in use.

The other elements of the six-point working definition of *business model* seem to fit well enough with IP-based business models. Focusing on the key aspects of the offering, developing the value chain and the surrounding value network, and specifying the payment mechanisms all seem to carry over to the IP-based space for business models—provided that the focal organization figures out a target market and a compelling value proposition. Each IP-based company is highly focused on a specific activity. Yet, each of these pure-play IP-based companies ultimately depends on one or more licensees developing business models with the IP that satisfy these requirements. These pure-play companies' own business models must, at a minimum, make room for their licensees to be economically successful with the IP in order for the companies to sustain the high-margin, asset-light financial flows that they seek to obtain from the licensees. In brief, a pure-play IP-based business model must be crafted so as to sustain the emergence of viable business models among the licensees or clients of its business model.

IP-AWARE OPEN BUSINESS MODELS

As business models become more open, there will be more opportunities to inject external ideas and technology into one's business model. Open business models also will make greater use of external parties to commercialize internal ideas and technology via others' business models. In this chapter, we have seen IP-based companies that have developed business models that rely exclusively on external parties to commercialize the IP. These companies may seek to buy your unwanted IP. These companies may seek to license IP to you. While these pure-play models vary in their focus, each successful model nonetheless must align itself with the business models of the commercializing partners as well. This will create IP-aware business models.

The pure-play IP business models offer new ways for people and firms to find markets for their inventions as suppliers, and to tap those markets for new products and services as customers. Since these models require less up-front cash and fewer assets, they can promote greater entry into the upstream market for ideas that otherwise might be blocked. As we have seen, these models still require significant investments of time and resources to make them sustainable over time. Given the success that some of these models have had to date, though, we are likely to see more of them in the future.

These IP players offer yet another dimension to open business models. These players are a potential reservoir of useful knowledge that can solve important problems in the market. They amount to another channel for ideas to get to the market. As Open Innovation–minded firms seek out new and differentiated technologies and ideas to compete in the market, this class of companies will become another important source to monitor, engage, and manage.

8

Getting from Here to There

B y this point in the book, you may be persuaded that your business model needs to open up and incorporate the benefits of intermediate markets for innovation and IP, while managing the risks associated with those markets. However, there is undoubtedly a nagging voice in the back of your mind that wonders whether, and how, to get from the business model you have today to the more open business model to which you aspire.

One path to get to there involves a review of the business model framework in chapter 5. Review your own business model, and determine where your model currently fits. Then carefully examine the next type's model, and make a list of the changes that will be required to achieve that next type of development for your business model. To help you assess this, there are some diagnostic questions at the end of that same chapter, in table 5-2.

That assessment, however, only tells you *what* to do next. You still have to figure out *how* to do that next set of tasks. This final chapter is intended to recount the experiences of three rather different organizations and how they managed to make significant changes that opened up their business models. While your situation will undoubtedly differ from those faced by these companies, there are likely to be some aspects of their experience that will be relevant for your own journey.

I have chosen to study IBM, Procter & Gamble, and Air Products. They operate in three very different industries, with very different technologies and products. Each used to function with a very internally focused, closed business model. Each has moved to a far more open business model. And each encountered significant challenges along the way.

Notably, these very different journeys had some common elements that are likely to arise in thinking through your own transformation of your business model. Each journey began with a "shock to the system," a significant event that helped the company alter its conception of its business. Equally importantly, the shock convinced others in the company that "business as usual" was no longer likely to be effective. This laid essential groundwork for later change.

The second step on the path was to try some experiments with the business model. The business model framework might suggest certain kinds of experiments (such as how to enlist more external technologies or partners in your own business model, or how to unleash unused internal ideas and take them outside) that each company had little information about initially. None of the three companies knew in advance exactly what was required to greatly improve its business model. To learn more, each company had to launch exploratory investigations, many of which ultimately proved unhelpful.

The third step was to evaluate the results of these experiments and recognize a promising outcome that might be a harbinger of a better business model. This wasn't just recognizing that something worked in a pilot study; it also required each company to assess whether it would be likely to work at a much larger scale across the company.

In the fourth and final step, these companies started to scale up the results of the early experiments. If that process went well, the company would then announce the new business model, both within the company and to external stakeholders. In this phase, the company also had to figure out whether and how to sustain the current (soon to be "old") business model, while transitioning to the new model.

I provide an account of each firm's experience, based on interviews I have held with numerous executives at each firm (as well as external articles published on each firm). After each account, I compare and contrast these characteristics along each of the four steps just noted and develop the implications of each step for those who wish to embark on their own initiative to open up their business models.

IBM'S OPEN BUSINESS MODEL:
AN OLD DOG LEARNS NEW TRICKS

Readers of a certain age will likely recall that IBM was a very different company before 1993. For those who do not recall, or are not old enough to know, IBM in the 1960s and 1970s was regarded much as Microsoft is regarded today: a very large, enormously successful, extremely well-managed company that effectively was a monopoly—or nearly so. IBM not only dominated the mainframe computer market (with 70 percent of that segment's sales and 90 percent of its profits), it also dominated the associated hardware and software products within that segment, such as monitors, printers, disk drives, tape drives, relational databases, FORTRAN, and COBOL. IBM was so successful as a company, and so dominant in its markets, that the U.S. Department of Justice filed suit against it in 1969 for antitrust reasons, alleging anticompetitive behavior by IBM.

Within its mainframe business, IBM was squarely positioned in type 3 of the business model framework. IBM had enormous scale and deep vertical integration, enabling it to lead the market in many technologies. However, it was almost entirely closed in terms of its ability and willingness to use any external technologies. IBM not only invented, developed, and manufactured its own products; it sold them, serviced them, and financed them through its own organization. From the basic materials all the way to the on-site service, everything was provided by Big Blue.

It is also well known that IBM developed many, many important technologies that never seemed to get to the market—until other companies appropriated the technology and did it themselves. Oracle commercialized the relational database technology that IBM invented but failed to pursue internally (until long after Oracle developed it). Sun Microsystems and other workstation companies marketed reduced instruction set computing (RISC) processors against products from IBM, though IBM had RISC processors in its labs long before these competitors. And in the PC business, Intel and Microsoft appropriated most of the value created by the IBM PC architecture.

IBM's Closed Model in Crisis

In 1992, IBM's type 3 business model reached a financial crisis: the mainframe market had matured; IBM's PC market shares were in terminal decline; the server and workstation businesses were far behind the market leaders; and the software business was in disarray. In December 1992, IBM announced its first major layoffs in its corporate history, and what was then the largest loss in U.S. corporate history: $5 billion.[1] Soon after this announcement, IBM fired its CEO and brought in the first outside CEO the company had ever had, Lou Gerstner.

Gerstner's arrival at IBM, and the subsequent changes to IBM's business model under his direction, have been widely studied.[2] However, the process that IBM went through to get to its new business model has not been widely reported.

In the beginning of its transformation, IBM decided that its overhead structure was too bloated for the amount of revenue coming into the company. As noted earlier, this led to an extraordinary layoff and the writing off of many corporate assets, which were the principal elements of that $5 billion quarterly loss. Among the early casualties were many members of IBM's R&D staff. This was radical, short-term surgery, designed to stop the financial bleeding. The days of lifelong employment at IBM were over, and every surviving IBM employee knew it.

The Hunt for New Revenues

Once the bleeding was stopped, many organizations within IBM began a fervent hunt for new revenue sources. Since some of IBM's core businesses were maturing or even declining, IBM managers knew that more cuts would be coming soon if new revenue sources could not be obtained. Three areas within IBM that illustrate this hunt for new revenues were the semiconductor business within IBM, its IP licensing and management business, and its open source software initiative.

Within the semiconductor business, Gerstner mandated that the business become cash flow positive. Before this mandate, IBM had invested in the semiconductor business as a strategic business that provided technological advantages and differentiation to its workstation, server, and mainframe businesses, which all used IBM's chips. On a standalone basis, however, the semiconductor business was losing significant sums of money. To meet the Gerstner mandate, new semiconductor revenues were urgently needed.

This search for revenues prompted some experimentation with the business side of IBM's semiconductor business. One experiment was to offer IBM's own semiconductor lines to act as a foundry for other companies' products. This had the positive benefit of bringing in new revenue, and it helped increase the equipment utilization rate of IBM's semiconductor process equipment and facilities. This meant that IBM's internal fixed costs were now spread over higher unit volumes, lowering the cost of IBM's chips to its internal customers within other parts of IBM. A related experiment was to create a research alliance within semiconductor process development to share the high costs and significant risks of pioneering leading-edge semiconductor processes. IBM was now breaking even, or maybe even making a little money, in an area that before was costing the company tens of millions of dollars in losses each year—and was sharing the risks of this effort with its partners.[3]

A second area of initiative within IBM to generate funds has proved to be even more innovative. IBM's need to generate profits in its semiconductor business caused it to rethink its whole approach to managing its IP, especially its patents and technology. According to Jerry Rosenthal, IBM's vice president of intellectual property in the 1990s:

> Prior to Gerstner, we only licensed our patents. An impetus for us to broaden that policy came from watching the amazing pace at which Korean firms caught up in semiconductors. They came from nowhere to become major players. They didn't just want our patents, they wanted our technology— our know-how, our trade secrets—to tell them *how* to use our patented technologies. And while we wouldn't give this

to them, many Japanese companies would and did. As a result, Lou Gerstner agreed to open up our licensing to include licensing our technology as well as our patents after seeing this.[4]

IBM's know-how and trade secrets were important ingredients for IBM's licensing, since they helped licensees figure out how to make the IP useful in the semiconductor business. As with Qualcomm in the previous chapter, even straight licensing often requires practical industry experience in how best to use a technology.

As one might imagine, there was some resistance within IBM to the decision on licensing its technology. From one part of the organization—R&D—some felt that IBM was departing from its historical approach to preserving freedom of action. From other quarters—the businesses—some managers felt that IBM was practically giving away some of its best technology to competitors of those businesses. As Catherine Lasser, IBM's VP for solutions engineering put it, "Often the initial feeling of the businesses was, 'Why should we let it go and potentially compete with ourselves?'"[5]

If IBM's financial condition had been very strong, this internal resistance might have carried the day, or at least delayed the change in these policies. But the overriding consideration at this time was to find more revenues, and the businesses were having a hard time generating new revenues on their own. IBM's financial distress meant that the alternative to licensing out IP and technology could be another large round of layoffs.

A third set of experiments were being conducted in the software area. In the operating systems market, IBM was losing share in the 1990s to both Unix and Windows NT. Moreover, as these systems were exploding in volume, IBM was conscious that these products had strategic importance beyond their revenues—they were key control points that would determine the direction of new technologies and architectures for enterprise computing. And enterprise computing was IBM's bread and butter. If IBM couldn't get back out in front and help steer these control points, the company risked losing its position as a leader in one of its core markets.

It was in this context that some IBM programmers and managers were evaluating the potential of the Linux operating system. Linux by itself would hardly solve IBM's revenue problems (since the code base was available to anyone on an open basis, it lacked the ability to generate revenues for IBM the way Windows NT had done for Microsoft). However, Linux did offer IBM the opportunity to get back into a leadership position in software operating systems, if IBM could find other ways to make money from it. As Joel Cawley, IBM's VP for corporate strategy, put it, "We needed some horse to ride to grow our own business. We were watching Linux, and though the business was small [2 percent share] and only ran on Intel processors, it was growing fast, and it had already attracted a large following."

The software business was building a different kind of business model out of IBM's experiences. Instead of the proprietary business model that IBM had long practiced in the software business (FORTRAN, COBOL, DB2, AIX, to name a few of its leading proprietary products), IBM's software business began to embrace Linux and to construct its own business model around the Linux code.

This required some major internal changes within IBM and also necessitated IBM to influence the opinions of many outsiders who were skeptical about working with IBM. It wasn't easy. As Jerry Stallings, IBM's current VP of IP and strategy described it, "IBM's reputation was a big, sometimes arrogant, company that takes over whatever it gets involved in. We had to learn how to collaborate."

IBM's Open Source Business Model

The changes and the work to achieve them have paid off. Cawley outlines the logic of the new business model for Linux within IBM:

> I have long observed that it takes $500 million to create and sustain a commercially viable OS [operating system]. Today we spend about $100 million on Linux development each year. About $50 million of that is spent on basic improvements to Linux, how to make it more reliable. The other $50 million is spent on things that IBM needs, like special drivers for particular hardware or software to connect with it. We asked the Open Source Development Lab to estimate how much other commercial development spending was being done on Linux. This didn't count any university or individual work,

just other companies like us. They told us the number was $800 [million] to $900 million a year, and that the mix [of basic versus specific needs] was close to fifty-fifty. So that $500 million expense for a viable OS [operating system] is there now for Linux as well [counting only the basic portion, not the specific portion]. And we only pay $100 million toward that. So you can see even from a very narrow accounting view that this is a good business investment for us.

So IBM has regained an important control point for the enterprise computing market, and it has done so through a creative, collaborative process that shares the costs with many other companies. According to the business model framework, this is a powerful example of a type 6 business model.

IBM's success with Linux, however, didn't come for free. It came largely at the expense of its own version of the Unix operating system, called AIX. While AIX was not growing much in the late 1990s, it was profitable. Although Linux was growing much faster, it was much harder to see how Linux would ever be profitable to IBM. There were protracted disagreements within IBM over this issue. In the end, however, the ability of Linux to recapture the strategic high ground (those control points in enterprise computing architectures) proved to be decisive. In 2001, the internal victory of Linux was announced to the world with Lou Gerstner's proclamation that IBM would spend $1 billion on open source software that year.[6]

The open source initiative that received IBM's pledge to spend $1 billion has been boosted by recent actions by IBM as well. One recent action is the decision by IBM to donate five hundred of its software patents to the open source community. This donation was intended to increase the intellectual commons available for the further development of open source software. It will likely be followed by additional donations by IBM in the future. The donation has already elicited copycat donations from Computer Associates and Sun Microsystems, while Nokia has announced that it will not enforce its patents against open source developers. Relatedly, IBM has joined with Intel, Dell, HP, and Novell, among others, to indemnify Linux users against lawsuits for IP infringement. IBM has also taken the lead in

defending Linux customers and developers from Santa Cruz Operation's lawsuit alleging copyright infringement.[7]

Getting Internal IBM Technologies Out

The solutions engineering group within IBM has recently emerged
with a different charter: find external homes for the myriad technologies within IBM that are not taken up by IBM businesses. Lasser is
VP of this group. As she explains her group's role, she says:

> Our task is to figure out what to do with the output that re-
> search creates that is not used in products internally. Generally,
> 80 percent of our output goes into our core businesses, but
> 20 percent doesn't. After all, we are a research organization.
> That 20 percent is what we focus on . . . My research col-
> leagues get tired of me saying it, but I tell them constantly
> that, "it's not just the technology; it's the business model."

Her group must research and develop alternative business models in the course of placing underused ideas and technologies externally. So her team includes business development people, contracts
executives, and legal experts as part of performing this function.
Moving technology out of IBM frees up resources and management
to look at new opportunities, and it enables IBM to make some profit
out of ideas that otherwise might go unused.

IBM Confronts Imitations of Its Own Success

IBM's successful practices in creating more value from its IP have
been widely imitated. Ironically, the new focus on creating value from
IP has launched a group of companies that own but do not practice
the IP and that seek to extract value from other companies that practice infringing IP. As Jesse Abzug, an IP manager inside IBM's research organization, explains, "During the 1990s, we trained many
other companies on how to manage IP. Ironically, we are now being
attacked more than ever by small assertion firms who learned many
of the ideas from studying us. Some of them are referred to as 'patent

trolls'. . . Kevin Rivette is heading up a task force that is looking at the problem of what to do about these trolls."[8] It seems that IBM needs to innovate again to develop new responses to the opportunities and challenges of creating, managing, donating, and profiting from its IP.

IBM Emphasizes Services

Under Gerstner's management, IBM began to look for the businesses where things were working and continued to exit businesses where IBM was clearly not succeeding (such as the OS/2 business). One insight that Gerstner brought with him was the need for IBM to help its customers integrate and manage all of the complexities associated with leading-edge information technology. This grew into IBM Global Services, which is now more than half of the company's revenues.

IBM is also embarking on a series of initiatives in the services sciences, management, and engineering, which promise to bring IBM's innovation capabilities to bear on what is now the largest, and most rapidly growing, part of its business. Going forward, the company is making a major new thrust into ways of innovating its services development and delivery to its markets. As Paul Horn states, "If you talk to our CEO [who is now Sam Palmisano], he wants IBM to be *the* innovation company, not merely *an* innovative company. This means we have to help our customers be successful and help them to innovate their own businesses."

PROCTER & GAMBLE: WE INVENTED "NOT INVENTED HERE"

P&G is at the forefront of using both internal and external innovation to drive toward a new business model. But this $56 billion organization did not get to its current position easily. The roots of its powerful transformation came from dealing with a serious crisis that the company faced only a few short years ago.

Back in 1990, P&G aspired to double its sales by 2000. However, when the year 2000 came around, P&G's sales fell short of this goal by $10 billion. This became known within the company as the

"$10 billion growth gap" and prompted a companywide initiative called Organization 2005 to try to close that gap. Wall Street was unhappy too. As Jeff Hamner, P&G's VP for corporate R&D, remembered, "P&G used to be dinged by the analysts because its R&D expenses were higher as a percent of sales than our competitors in the 1990s. We were felt to be less productive."[9]

P&G Misses Its Earnings

Durk Jager, who was CEO from 1998 through the middle of 2000, started a number of initiatives to restore growth to P&G. While many of these initiatives were very helpful in rethinking P&G's business, they created significant disruptions in the day-to-day running of P&G's business. The new initiatives took time to bear fruit, and the current business began to underperform—it was a deadly combination. During 1999 and the first part of 2000, P&G missed a number of consecutive quarterly earnings forecasts. This caused P&G's stock to plunge, from more than $110 per share in January 2000 to half that amount by May 2000. On June 8, 2000, P&G announced a change in management. Jager departed, and A. G. Lafley, who was running the North American beauty care business, was brought in to replace him.

In studying the causes of the growth gap, P&G realized that the problem was not in the performance of its current brands. These were leaders in their respective markets, holding the number one or number two position in market share. But these markets were maturing, which limited the growth of the current businesses to the demographic growth of the U.S., European, and Japanese populations. The source of the gap was that P&G was not developing very many new brands. For P&G to grow faster, it needed to increase its rate of new brand introductions. And it had to hit its numbers in its current business in the meantime.

Closing the Growth Gap: Connect and Develop

Working with Gil Cloyd, P&G's CTO, Lafley put a stake in the ground: to get P&G to accelerate its growth by opening up its innovation process to external sources of technology. Under an initiative called

Connect and Develop, Lafley proclaimed that in five years' time, P&G would receive 50 percent of its ideas from external sources. Many of these new ideas would come in the form of new brands, such as the SpinBrush and Swiffer brands, which were two early successes from this initiative.

This was not an easy transition for P&G's internal staff to make. P&G businesses sought to protect its technologies so that other businesses, including potential competitors, could not use them. R&D staff wanted to advance internal P&G technologies and didn't see any value in working with external technologies. "We invented 'Not Invented Here,'" recalled Jeff Weedman, vice president of P&G's external business development.

To the company's credit, these tumultuous events transformed a large, internally focused company into one with a much more open approach to innovation. Instead of terminating the Connect and Develop initiative, P&G recommitted itself to the approach. P&G took a team of R&D people under the leadership of Larry Huston and gave them the charter to realize Lafley's goal of 50 percent external inputs into the company's innovation process within five years. As Huston pointed out, "Our current, internal innovation business model was not sustainable. We had to make a discontinuous change. I set a goal with my boss to double our innovation capacity at no increase in costs. We started with roughly 8,200 people: 7,500 inside, 400 with suppliers, and around 300 virtual or external people. Now we are at about 16,500: we still have 7,500 internally, but now we have 2,000 with suppliers and 7,000 virtual and extended partners. I see myself as being in charge of those 9,000 people."[10]

Licensing Out P&G Technologies

Another change was paying greater attention to external licensing of P&G technologies. In the beginning, recalled Martha Depenbrock, an executive in P&G's external business development group, "We found that many of the patents we didn't want others didn't want either. And the really good stuff was being held back by the businesses. This is when the 3/5 program was created. Three years after the product [was] shipping into the market, or five years after the patent [was] issued, the patent would be made available to others . . . Nothing was untouchable."

Simply getting patents available for license, though, was only the first step. "Our next problem was to find those companies that were interested," said Depenbrock. "In the beginning, we followed the *Casablanca* approach, as in the part of that movie where the inspector says, 'Round up the usual suspects.' This had limited benefits . . . So then we became more strategic about what we said and where we said it . . . All of this is to try to build our own network on the business side, to complement our technical network being created on the Connect and Develop side."

Mike Hock, P&G manager for IP and know-how licensing, has led P&G's efforts to apply this new thinking to other types of P&G assets besides patents. "Know-how is all the IP stuff that ain't patented," he commented. "We think that we have applied the Open Innovation concept to the world of know-how. One example of this is a manufacturing process control algorithm that we developed in our manufacturing. While the algorithm was good, it took a lot of expert time to train people on how to use it and how to apply it to their operations. We created an exclusive license with another company who took it out and further developed it. They made it more reproducible and reduced the amount of expertise required to use it successfully."

The company that bought the license was BearingPoint, Inc. (formerly KPMG Consulting), which marketed the technology under the name PowerFactor. Since the agreement in 2002, PowerFactor has delivered savings of more than $150 million to BearingPoint clients.[11]

Hock reported that something interesting happened when this internal technology met with success on the outside. The fact that outside companies were licensing this algorithm made the not-invented-here-minded internal manufacturing staff sit up and take notice. Hock reported, "We eventually reimported the product, and our internal rate of adoption went up by a factor of three . . . This was due both to the value added by the further development, and by the external validation of having something that was succeeding on the outside.

Sometimes, we are more attracted to something that has been used outside than we are to our own internal stuff."

Large, deep companies with thousands of employees spread across many businesses often have internal ideas that don't get the attention they deserve. Ironically, the fastest way to call attention to an unused internal technology within such a company may be to sell it to others on the outside first.

Openly Partnering for Growth

Another superb example of a more open business model within P&G is its new equity joint venture with one of its oldest competitors, Clorox. Clorox had acquired the Glad brand from SC Johnson, and the brand had no technology to create follow-on products that would sustain its differentiation in the market.[12] Without some new technology to differentiate itself, the Glad brand faced the risk of commoditization.

As it turned out, P&G had two internal concepts that would create meaningful differentiation: its Press'n Seal and ForceFlex technologies. These ideas had even been taken into a local market test, where they rose to number one in the category of baggies and garbage can liners. However, at this time in 2001 and 2002, P&G was going through a financial crisis and lacked the resources to launch a new brand in this category.

The old P&G might have sat on these technologies and done nothing with them. The new P&G created an auction process to see how to get the most value from its technologies, whether they were deployed inside or outside the company. In addition to considering an internal business to commercialize its two technologies, it sought out potential partners and soon identified Clorox as a good candidate. When P&G ran the numbers on the alternative ways to commercialize its technology, the Clorox deal trumped the internal approach, in part because this particular deal didn't require P&G to spend money up front to gain distribution for the new brand.

After months of careful investigation and planning, the two companies created a joint venture in January 2003 in which Clorox held 80 percent and P&G received 10 percent and held an option to obtain another 10 percent. The venture performed so well that P&G

did exercise its additional 10 percent option in January 2005, at the price of $133 million.[13]

Subsequently, Clorox has approached P&G to take some of its brands into other parts of the world, such as Japan, where P&G has strong distribution and Clorox does not. This follow-on business opportunity was never envisioned when the two companies first started talking, nor was it part of the financial modeling that led P&G to work with Clorox. Yet this opportunity fits a larger pattern that Weedman refers to as Weedman's Corollary to Moore's Law: "The second deal takes half the time of the first deal. The third deal takes a third of the time, and so on. And the subsequent deals are not only faster; they tend to be more profitable." The implication of Weedman's observation is that you don't get to do the second or third deal unless you have done the first one.

Weedman's Corollary means that P&G doesn't maximize the value of the initial transaction because it wants the opportunity to pursue later deals. Gil Cloyd, P&G's CTO, puts it this way: "We are getting very interested in how to build better collaborations with others. We have used customer creation teams for many years with key customers, in some cases going back to the 1980s. This is not easy to do, however, because there are tensions that arise in who owns what IP. The best partnerships require both sides to make accommodations in order for the potential from the collaboration to be realized."

P&G is playing for the long run and now approaches its partnerships in more open and creative ways. This has altered its view of how to achieve competitive advantage.

The P&G Recipe for Competitive Advantage

In the process of managing relationships like the one with Clorox, Weedman has developed a far more nuanced view of competitive advantage. "There are many kinds of competitive advantage," he says. "The original view was: I have got it, and you don't. Then there is the view, that I have got it, you have got it, but I have it cheaper. Then there is I have got it, you have got it, but I got it first. Then there is I have got it, you have got it from me, so I make money when I sell it, and I make money when you sell it."

P&G feels that this last form of competitive advantage is the most sustainable form in the current environment. It fits very well with the type 5 business model in the business model framework. As Steve Baggott, director of external business development at P&G commented, "If you don't license, chances are very good that someone else has a very good technology too. It's rare that you're the only game in town. So do you want to participate in the licensing revenues or not?" Indeed, P&G welcomes the notion of having its competitors use its technology. This is because competitors that employ P&G concepts are not striving for newer—or even better—concepts in their own labs. They become "happy followers," instead of "threatening leaders." Happy followers allow P&G to sustain industry leadership and help defray P&G's costs of doing so.

Becoming the Innovation Partner of Choice

Where will P&G go from here? Its modest ambition is to become the partner of choice for inventors and companies with good ideas that are looking for a way to get to consumer markets. As Weedman points out, this changes the way that P&G must behave in working with others. "How do we make ourselves a partner of choice if you bring us an idea? The small things actually matter here. If you send us something, we will acknowledge it. And we will respond to you as well. Ninety-nine percent of the time we are not interested, but we will say, 'Please call again.' And if it looks good but not for us, we will refer it on, even to a competitor. Why? So that next time the inventor has an idea, he will contact us first."

P&G also works the other side of the street. In discussions with Wal-Mart, Weedman asks Wal-Mart to send on promising ideas that it receives but that are not ready for the volumes and logistics that it requires. Weedman says of Wal-Mart, "I invite them to help us find their next product from us. Many times, there is a very good idea but the company can't pull off the volume production needed to meet Wal-Mart's requirements. We can."

This is a logical destination for one of the world's preeminent consumer products companies: don't rely only on your own staff to dig up great opportunities. Instead, shift from "push" marketing, where you have to uncover the opportunities, to "pull" marketing, where people come to you. P&G's CEO A. G. Lafley summarized the opportunity for P&G: "I honestly believe that we can increase our innovative capacity fivefold by collaborating more effectively with external partners. I want to create an environment in which there is an open market for ideas, for capital, and for talent. More specifically, an environment in which big ideas can attract the capital and talent they need on the strength of the idea itself no matter where it comes from."[14]

Become known as the partner of choice, and use that reputation to attract promising ideas to you. If IBM aspires to be the innovation company in the enterprise market, P&G seeks to be the innovation partner for the consumer world. If P&G is able to pull this off, it would clearly be in a type 6 business model.

AIR PRODUCTS: SMALL IS BEAUTIFUL

Air Products is in an industry that at first blush seems far distant from the cutting edge of innovation: industrial chemicals. Many of Air Products' offerings indeed are very mature industrial chemicals. Yet this $7.4 billion company has quietly refashioned itself into quite a leader in innovation. In contrast to IBM and P&G, Air Products is pretty low key in publicizing its own success. Yet its approach to a more open business model holds many insights for companies in any industry.

It is no exaggeration to say that a decade ago, Air Products was a very old-fashioned company when it came to managing innovation and IP. According to Gus Orphanides, a twenty-five-year veteran of the company now responsible for its intellectual asset management and licensing, "Our approach to licensing and IP was defensive. We patented to defend our market positions and our business assets so that no one else could practice our technologies."[15] One of the executives in the intellectual asset management team for Air Products' electronics group concurred: "I was a member of the Patent Technology Council in those days, and we never could get business people to show up to our meetings . . . We would review about forty inventions in two to three hours, so there were only a few minutes per idea . . .

We would typically get the big stuff approved that involved significant internal commercial use done quickly. But all the other stuff just languished." At this time, there was very little external licensing of any of Air Products' technologies. This was a type 3 business model.

So what changed? What got this stodgy company to revitalize its business model? One key impetus was a merger that almost happened but was stopped at the last minute. Air Products and a competitor, Air Liquide, were jointly going to acquire British Oxygen, with some parts of British Oxygen going to each acquiring company. It fell to Miles Drake, the then-new CTO of Air Products, to figure out how Air Products was going to integrate its pieces of British Oxygen. To Drake's credit, he went far beyond drawing boxes on an organization chart.

Drake began to think about how Air Products could innovate differently with these new parts to work with. For example, what should Air Products do with British Oxygen's products? Its patents? Its other intellectual property? How could all these assets create more value for Air Products?

Unleashing Intellectual Assets

As it turned out, the merger never went through. But Drake's out-of-the box thinking stuck in his head, and he realized that he didn't need a merger to implement his ideas for a new way to innovate and compete. Air Products had plenty of its own intellectual assets, and Drake knew that little was being done to create value from them. In meeting with his management, he put forward the case for unleashing these assets. He suggested that the corporate executive committee (CEC) adopt a policy that "all intellectual assets are corporate assets and should be managed for maximum shareholder value." The CEC changed only one word in Drake's proposal. The word *should* was replaced by *shall*, thereby strengthening the proposal and showing the whole company that the CEC was firmly behind this change.

This is where John Tao, a thirty-year veteran of Air Products, began to change the company's approach to licensing technology. At first, he simply asked the CEC for six months to benchmark how other companies were managing their intellectual assets. After six months of research, he then began to develop an outlicensing program for Air Products. As Tao explains, he started small: "I didn't ask for large amounts of money on purpose. I thought that if I requested a lot of money before we knew what we were doing, I would be painting a target on the program and make it an easy target for some future cost-cutting meeting . . . A colleague of mine from Boeing took the reverse approach. He built a thirty-person licensing group right from the outset and set big targets of financial performance to justify that size group. When tough times came, that group was disbanded, and many of the thirty people had to leave the company." By starting small, Tao has been able to learn how best to license his technology without spending much money or attracting top management's attention before he was clear in his own mind how best to proceed. His approach moved the company from a type 3 business model to a type 4 model.

Partnering for Innovation

Air Products has also changed its process for developing technologies for its own business. It has shifted from a strategy of doing all the research and all the commercialization activity itself, to a model where it partners with others as part of commercializing new technologies. One example of this is the company's approach to nanotechnology. Air Products has developed some powerful ways to manipulate nanoscale particles—in nanoclays (for composite materials), nanoscale oxides and metals (for higher-performance materials), and carbon nanotubes (for semiconductor materials). But instead of doing this work on its own, Air Products has partnered with DuPont and a small German firm, Nanogate Technologies, to commercialize these products. According to Martha Collins, technology director for Air Products, "The keys to successful nano projects are alliances and partnerships in the spirit of open innovation."[16]

In the process of licensing out internal technologies or partnering with others, Air Products has also changed its approach to managing intellectual property. Under the leadership of Herb Klotz, the company has created a series of teams for managing the intellectual assets of each of its businesses. Each team is free to determine what technologies have the most potential value, regardless of whether they would be used inside Air Products or licensed to others. As Jeff Orens, in the advanced materials intellectual asset management team, noted, "Now our scientists are also excited. They think 'Whoa! Now I have a business reason to pursue this research and pursue a patent. It's not about protecting our old stuff anymore' . . . What I tell our scientists is that what we are protecting is not you or even your invention. What we are protecting is the business model." More specifically, this integration of internal and external technologies and opportunities is indicative of a type 5 business model. (Note, though, this thought process is still in its early stages across all of Air Products.)

Large Scale Vortex: An Innovative Model for an Innovative Technology

An example of this open approach to intellectual asset management is the Large Scale Vortex (LSV) burner technology that Air Products has developed. This technology reduces nitrous oxide (NOx) emissions from industrial combustion in plants making chemicals like hydrogen, methanol, and ammonia. The burner is part of the air pathway for chemicals to exit the plant to the outside. By burning the extra effluents, it reduces NOx emissions, improving air quality from the plant. The irony is that this technology had been inside Air Products for some time but was not very popular within the company. "Once you have built a $100 million plant, you want to use it for a while before you start refitting it," explained Orens. "In fact, it is the external interest we have received in LSV that is causing our internal businesses to sit up and take notice of it." This is reminiscent of P&G's experience with its manufacturing algorithm.

Orphanides described how Air Products then approached the pricing for its LSV technology: "We were thoughtful about developing the business model around this technology . . . We calculated the value that our solution created when incorporated at the customer's site. We then discussed how much of the value to give to the end users, our direct customers, and ourselves. We knew that there had to be a positive for everyone for the business model to work." This is very sophisticated pricing. On the one hand, the technology is being priced on value, not its cost. On the other hand, the value pricing is determined by the need to provide value in excess of the price throughout the value chain.

Sustaining and Expanding the Model

With Drake's and Tao's guidance, and Klotz's management of the various intellectual asset teams, Air Products has crafted a sustainable process that is changing the entire company's business model into a type 4 model. The outlicensing programs are projected to double in revenues in 2006. And these are very high margin revenues for an overall commodity business. The internal businesses are getting credit for these new revenues and profits in their own business. They are paying closer attention to other new technologies being developed within Air Products. And the CEC is finding that the program is delivering these results off of a modest budget. Now the CEC is asking what more could be done, a positive sign for the future of innovation at Air Products. And the intellectual asset teams now enjoy the credibility within the company that comes with delivering a positive track record of results that exceed the expectations set when the program was started. As further successes are achieved, Air Products will likely progress into a type 5 model across many of its businesses, which integrates internal and external technologies both in its internal R&D activities and in its external licensing and partnership activities.

IMPLICATIONS FOR OTHER COMPANIES SEEKING TO OPEN UP THEIR BUSINESS MODELS

While each company's experience in transforming its business model is unique, there are some important parallels that can be seen in all of them. These parallels offer some insight and direction for other companies that want to examine how they might get to a more open business model themselves. These companies' experiences also document some likely barriers that companies may encounter as they strive to do this.

As noted at the beginning of this chapter, these companies' journeys can be examined in four distinct phases, phases that likely will characterize any attempts by others to follow in their path. These phases are: (1) a shock to the system; (2) searching and experimenting to find new revenue sources; (3) identifying successful business models; and (4) scaling up and adjusting the model. I will discuss each phase in turn.

Phase 1: Look for a Shock to the System

Each of the companies began their journey with a shock to the status quo. In the case of IBM, the shock was so severe that the company was nearly broken up. It is not surprising that IBM chose to go outside for its CEO in 1993, because its board no doubt realized that significant cost cutting was necessary and that an outsider might be in a better position to perform the surgery. After all, these cuts meant the end of the near lifetime employment policies of IBM and shattered what had been a core value of the company.

In the case of P&G, the company's stock fell to half its original price in four months' time, and a new CEO was brought in. While P&G's situation was not as severe as IBM's crisis, it was clear to everyone in the company that things could not remain the same. However, P&G did not need to bring in an outsider; it brought in a leader who knew the company very well but was not a prisoner to its old habits.

Air Products did not face the brutal financial adjustments that IBM and P&G endured. However, the potential partial acquisition of British Oxygen triggered a deep examination of how the company did business and how it might position itself to achieve better results in a mature commodity business.

This pattern is sobering. It implies that simply reading a book about more open business models is unlikely to get you from where you are to where you want to go. There are tremendous sources of inertia inside all companies, and successful companies develop even greater inertia because of their previous success. A significant shock to the system seems to be required to overcome the inertia that develops inside large, successful companies.

What to do? If you are heading into a period of crisis, there may be opportunity amid all the pain and difficulty of coping with the crisis. Such conditions do allow managers in companies to suspend their usual defense of the status quo, and at least some leaders within the organization may be eager for a path that might lead to improved performance. Opening your business model may be such a path.

If you are not heading into a crisis, all three companies show that there needs to be clear commitment and support from the top of the organization. P&G is the leader in this dimension, as Lafley's explicit involvement strongly endorsed the Connect and Develop approach to innovation. Gerstner also played a key role at IBM. Though he didn't mandate or endorse the solution early on, he defined the conditions and ground rules for it, such as his mandate that each business become profitable. When the inevitable organizational resistance comes, such as the "not invented here" mentality, support from the top can be critical to overcoming that resistance.

If top management support is not forthcoming, Air Products' approach provides a more subtle way to effect change. Start small, to stay off the corporate radar screen, and give yourself time to learn. And keep the expectations (and the budget) modest at this early stage. Wait until you have some clear, demonstrable evidence of your approach before calling attention to yourself.

Shocks to the status quo unfreeze the habits of the existing business and the existing business model. However, shocks alone are not enough. To continue to change, there must be evidence that supports the change and shows that the company is heading in the right direction. This brings us to the next phase.

Phase 2: Construct Experiments to Search for New Revenue Sources

Many companies encounter shocks and make drastic adjustments (such as cutting back on people and expenses) to deal with them. Fewer of those companies engage in the breadth of experimentation that IBM, P&G, and Air Products did when they were searching for a new business model.[17] It takes courage and vision to experiment with new ideas during a time of financial difficulty. Yet without these experiments, companies could easily fall into a vicious cycle of slowing business revenues, leading to head count and expense reductions, which trigger further business declines and lead to still more cuts.

One need only look at Ford and GM in the automotive business, whose market shares have been in a slow, inexorable retreat since the oil shortages of the 1970s, to see such a vicious cycle in action.[18]

Instead, companies that experiment with alternative sources of revenue and business value begin to collect critical information from the market about the potential value of some of their ideas, technologies, and markets through these experiments. The results of their efforts formed the seeds of the shift toward a new approach to their respective businesses. IBM, for example, began to offer its semiconductor facilities as a foundry and started to obtain significant sums for its patents and technology through external licensing. Not only did this help the semiconductor business remain profitable (and preserve the business within IBM), but it pointed the way for other parts of IBM to manage their own IP.

At P&G, the early successes of SpinBrush and Swiffer products, which were licensed in from other companies, provided proof within P&G that its Connect and Develop initiative could generate strong bottom-line results for the company. The later success with Clorox in the joint venture has shown P&G the power of collaborations not only in launching new products and brands but also in taking current partners' brands to international markets. These events have broadened P&G's sense of its business advantages and how best to leverage them.

At Air Products, the external interest in licensing its LSV technology not only provided new revenue to the company; the external interest also gave the technology greater credibility within the company. And the use of intellectual asset teams within Air Products has allowed the company to share successful internal practices across a diverse range of products and markets. The nanotechnology partnerships, by contrast, help share the risks and expenses of developing and marketing new materials with other parties outside the firm.

Testing and experimenting are critical inputs to the process of transforming the business model, for it is they that create the information needed to chart the course ahead. However, they too are insufficient to achieve lasting change unless a deeper understanding of a new business model emerges to connect the disparate results of individual tests into a larger, more coherent whole.

Phase 3: Recognize a New Business Model
from the Successful Experiments

Conducting experiments only yields value if a company is able and willing to act on the information generated by those experiments. While the impetus for the changes at IBM often came from the short-term imperatives of its individual businesses to get profitable, it is very much to IBM's credit that others in the company took note of these changes and began to consider whether there was an underlying logic that connected the results of these experiments. Larry Huston's early inlicensing of external products and Jeff Weedman's outlicensing experiences with some of P&G's internal technologies showed that there was money to be made as both a buyer and as a seller. But it was Gil Cloyd and A. G. Lafley who realized that there was a new logic underneath these individual practices, and that this new logic could transform P&G's business model and boost its overall growth rate.[19]

Air Products' experience to date is pointing that company in a new, more open direction for its own future business model. While the company continues to conduct significant advanced research, it now seeks out external university technology more assiduously. It also seeks business partners when it takes its ideas to market. The company is far more sensitive to the capital investment requirements that are needed for commercialization than it used to be. As Gus Orphanides remembered, "We used to be a huge CapX [capital expenditure] company, perhaps spending $1 billion a year for a $6 billion company. We started to ask ourselves, 'Are we getting enough of a return on our shareholders' capital?'"[20] Today, the company is more agile, and more creative, in working with others to share these costs.

In building a new business model, companies must also decide what to do about the current business model. Praising a new business model can inadvertently suggest that the current model is somehow obsolete. In all three of the companies profiled here, the current model continues to play an important role in the business of the firm. IBM's mainframe business continues to operate in a highly vertically integrated manner. P&G still develops its own brands and still invests a substantial amount in its own internal technologies. Air

Products still develops and supplies a great deal of its industrial gases on its own.

These companies must manage the coexistence of the current (say, a type 3) business model and the new business model (perhaps a type 4 or 5) within their organization. In each case, senior managers were tasked with managing the new business model, while other senior managers maintained responsibility for the current model. A delicate balance must be achieved between these two models. Indeed, when Durk Jager of P&G tried to push too many change initiatives at once, P&G did begin to change but lost the operational discipline to deliver the quarterly earnings numbers that P&G's investors expected. A. G. Lafley has restored this operational excellence while maintaining the change momentum. He found the balance between the current business model and the new model that eluded his predecessor.

It is interesting to note that not all parts of IBM have embraced these new initiatives with equal fervor. That is perhaps to be expected in a company of IBM's size. IBM's many businesses are large enough and autonomous enough that they may have different business model types. As competition, maturing markets, and globalization catch up to businesses with type 3 models, though, the new business models will help point the way toward continued growth.

Phase 4: Scale Up the Successes, and Proclaim the New Open Business Model

As successful experiments begin to point the way toward a new and more effective business model, the company must face a final phase in its transformation. In this final phase, the company must scale up the model to bring it into high volume across the company and the company's customers.

There are at least two essential elements in scaling up a new business model. First, the business model itself must be constructed or adjusted so that it can handle significant new volume. As we saw with innovation intermediaries InnoCentive and NineSigma in chapter 6, this can be a challenge. Business models that work when there are a small number of highly trained people to pay close attention can break down when new layers of administration are required to manage a much larger number of more general workers. If the processes cannot be more automated or standardized, they may not be able to handle large increases in activity without severely degrading the quality of the result. IBM faces this concern in its Global Services business. The skills of its services personnel differ from those of its product and technology personnel. IBM now needs fewer device physicists and polymer chemists, while it needs many more people who can translate customers' IT requirements into specific solutions that IBM can provide. The availability of such people is pacing IBM's growth in this area.

The second element is building a winning coalition within the company to gain the ability to roll out the model across the company. Building scale requires much more funding and much greater organizational commitment to the new model. These resources must come from somewhere. The rise of the new model can create "losers" in the organization, at whose expense some of those budgets may be obtained. There can be significant costs in persuading the company to make these changes in the face of resistance from the losers.

In IBM's case, the company had to adjust its business model in its IP management as one method among many to address this resistance. In the beginning, the royalties and revenues from patent licensing went to a corporate account within IBM. Now IBM directs the revenues and profits directly to the businesses associated with that IP. This reduces the friction that more open management of IP sometimes encounters from the business, which we saw in chapter 3. Now IBM businesses participate in the benefits, as well as the risks, of a more open IP management approach. P&G and Air Products also have adopted this "pass through" system, whereby the revenues earned from outlicensing now flow through the licensing group to the businesses within each company.

This is why John Tao's approach of starting small at Air Products makes so much sense. A small program requires fewer resources, less management attention, and less competition with other parts of the organization. Tao's program is now expanding, and there will be greater competition for resources in the future. Now, however, Tao's program has a track record and is bringing in new revenues, which also are shared with the associated business units. These new revenues dampen internal competition, since there is a bigger pie to share, and his successful track record to date gives him more clout in the discussions over how to divide the pie.

Getting scale with the new business model confers additional benefits that aren't achieved during the earlier phases of the change process. In IBM's case, for example, the company has repositioned itself from Big Blue, a maker of proprietary mainframes, to the New IBM, the champion of open source software. When Lou Gerstner publicly embraced Linux, that endorsement gave Linux a tremendous boost in IBM's core markets. As Jerry Rosenthal and Jerry Stallings note: "Lou's big proclamation back in 2001 . . . immediately legitimized Linux, and helped get it into the enterprise."[21] It also boosted IBM's presence in the enterprise by giving the company a new message to promote at a key control point in enterprise computing.

P&G is also benefiting from greater scale. Its growing confidence and commitment to Open Innovation is causing it to brand itself as the innovation partner of choice. This will attract more opportunities to P&G than it would receive if it had to seek out each one on its own. Of course, P&G must create processes that are able to handle a higher volume of activity to support its new brand positioning as the innovation partner of choice. But a rich menu of opportunities streaming into P&G will help it keep its competitors "happy followers" as P&G creates new business areas out of its technology and its innovation processes.

Air Products is not operating at the scale of IBM or P&G as of this writing. However, John Tao's group has carefully built up its credibility within the organization, enabling team members to exploit new, larger Open Innovation opportunities that previously would have been too ambitious for them to undertake. They are positioned to take their business model further in the future.

OPEN BUSINESS MODELS: PROFITING
FROM OPEN INNOVATION

IBM and P&G have timed their shift into a high-volume open business model very well. As this book has shown, there are growing intermediate markets for innovation. And these innovation opportunities are widely distributed, across universities, across countries, and across industries. Companies like IBM and P&G are well positioned to take advantage of this shift toward Open Innovation. Their competitors must either copy their approach or do what Clorox has done with its Glad brands and be a happy follower.

What about your organization? Can you afford to ignore the widely distributed innovation opportunities in your industry? Are you taking advantage of the emerging intermediate market for innovations? Could you add more value for your customers if you did so? Is your business model enabling you to do this, or is it getting in the way? And what would you do if your competitors started opening up their business models before you did?

Change in an organization is never easy. In chapters 1 and 2 we discussed some of the challenges that you will likely encounter as you open up your business model. Chapter 3 outlined the increasingly important role of IP in innovation. Chapter 4 provided a framework for how to think about the new context for Open Innovation, the increasingly important role of IP in innovation, and how to identify areas of opportunity, as well as risk.

Yet some changes are worth the costs, the time, and the headaches that they involve. In chapter 5, we saw a model that provides a framework for improving your business model. In chapters 6 and 7, we observed new firms that are paving the way for a world of more developed markets for innovations and their associated IP. We even saw some firms whose business models are aimed to exploit those firms that have not woken up to the new realities of managing innovation and intellectual property in their business models. In this chapter, we examined three different companies in three different industries that have made the journey toward a much more open business model and how they did it.

It is now your turn to decide where your business model is today, where you want it to be tomorrow, and what you are going to do to get from here to there. A world of opportunities, and risks, awaits those who dare to make the journey.

Notes

Chapter 1

1. For a thoughtful summary of the problems confronting innovating U.S. companies, see Richard Rosenbloom and William Spencer, *Engines of Innovation: U.S. Industrial R&D at the End of an Era* (Boston: Harvard Business School Press, 1996). More recent concerns have been ably voiced by the U.S. Council on Competitiveness in its report, "Innovate America: Thriving in a World of Challenge and Change" (Washington, DC: U.S. Council on Competitiveness, 2005). In Europe, see the Lisbon European Council conference report, calling on European governments to forge "the leading knowledge economy in the world" by 2010 (European Council, Brussels; see http://www.consilium .europa.eu/ueDocs/cms_Data/docs/pressData/en/ec/00100-r1.en0.htm for an English language version of the report.). Despite the concerns in the United States about its innovation system, the U.S. system (and to a lesser extent, Japan's system) is the benchmark against which Europe compared itself at the Lisbon conference.

2. See David Mock's book on Qualcomm, titled *The Qualcomm Equation* (New York: Amacom Books, 2005), for a very helpful, in-depth study of the company. Mock obtained excellent access to key leaders in the company, including those who were there at the beginning and have since retired.

3. A recent book by Gary Pisano, *Science Business* (Boston: Harvard Business School Press, 2006), shows that the biotechnology industry in which Genzyme participates has seen very few companies make a profit. Genzyme is one of only three companies (the others are Amgen and Genentech) of more than one hundred biotech firms that have demonstrated the ability to sustain profits in this treacherously difficult industry.

4. The story behind *Chicago* originated with Maurine Dallas Watkins, a Chicago-based journalist who covered the crime beat in Chicago when the murder of Walter Law occurred. Watkins reported the subsequent trial and

afterward wrote a play, *Chicago*, about those events. The play was performed on Broadway in 1926 and was made into a silent movie in 1927. It was revived in 1975 by Bob Fosse and revived again by Harvey Weinstein in 1997. The 2002 movie version of *Chicago* won six Academy Awards. Sources: "Chicago," Wikipedia, http://en.wikipedia.org/wiki/Maurine_Dallas_Watkins, and Richard Kromka, interview by author, March 16, 2004.

5. The ideas in this paragraph are inspired by the work of David Teece, Gary Pisano, and Amy Shuen, "Dynamic Capabilities and Strategic Management," *Strategic Management Journal* 18, no. 7 (1997): 509–533. This article is both a critique of business strategy, as well as a presentation of a concept called "dynamic capabilities" that describes how firms adapt their strategies to changing markets and technologies.

6. See Ashish Arora, Andrea Fosfuri, and Alfonso Gambardella, *The Economics of Innovation and Corporate Strategy* (Cambridge, MA: MIT Press, 2001), for a scholarly yet very accessible account of intermediate markets. The authors develop formal models for these markets and provide empirical accounts of how the models work in practice in the petroleum engineering construction and chemical businesses.

7. See David Teece's book *Managing Intellectual Capital* (Oxford, England: Oxford University Press, 2000), for an early account of the organizational implications of knowledge developing separately from the companies that use the knowledge. See also Joshua Gans, David Hsu, and Scott Stern, "When Does Start-Up Innovation Spur the Gale of Creative Destruction?" *RAND Journal of Economics* 33, no. 6 (2002): 571–586, for a more formal model of upstream technology suppliers that choose to supply to downstream customers or to forward-integrate to compete against those customers.

8. This judgment ignores the possibility that some companies need to obtain some patents as bargaining chips for use in disputes over patent infringement with other firms. This may be socially wasteful (since none of the patented technologies are used, but the firm owning the technology can prevent anyone else from using them), but these bargaining chips are of at least some value to shareholders who funded the R&D that created them in the first place. Brownyn Hall and Rosemarie Ham Ziedonis perform an insightful analysis of this behavior in the semiconductor industry; see "The Patent Paradox Revisited," *RAND Journal of Economics* 32, no. 1 (2001): 101–128.

However, one should not carry this point too far. While firms may need some bargaining chips for certain disputes in certain situations, this fact alone is unlikely to account for why the vast majority of patents go unused. Procter & Gamble, for one, quickly changed its management of its patented technologies once it realized how low its utilization rate was (as documented in

Nabil Sakkab, "Connect & Develop Complements Research & Develop at P&G," *Research-Technology Management* 45, no. 2 [2002]).

9. Patents issued in the United States now expire twenty years from the date of patent filing. This was changed from an earlier policy, in the late 1990s, of expiring seventeen years from the date of issuance. One interesting fact of many technology businesses today is that the patents live on long after the technology being covered has become obsolete.

10. IP protection provides some affirmative advantages as well. IP protection sometimes gives the company the ability to exclude other companies from using its technology against it in competing products. This delivers a little extra breathing room for the company in its markets. The companies that do the most R&D usually get most of the patents and other forms of IP that help them defend those businesses. Stock analysts have learned to inquire about the R&D investments of firms and became concerned when these were observed to decline. However, the degree of breathing room gained by IP protection has been shown to vary significantly by industry. See Wesley Cohen and Richard Levin, "Empirical Studies of Innovation and Market Structure," in *Handbook of Industrial Organization,* ed. Richard Schmalensee and Robert Willig (Amsterdam: North-Holland, 1989), ch. 18. In industries such as pharmaceuticals, IP protection is quite strong, while in most of the information technology industries, competitors can usually "invent around" the protected IP at an acceptable cost. One key difference in the strength of patent protection in pharmaceuticals versus information technology is the role of the Food and Drug Administration in regulating drug development. If a rival company invents around a company's drug patent, that rival still has to complete a series of clinical trials and regulatory reviews that take many years and many, many millions of dollars. Rivals that invent around information technology patents simply ship to market, with no regulatory review. Thus, patent protection has a much stronger effect in pharmaceuticals than it typically does in information technology.

11. For treatments of managing IP that focus exclusively on the value capture portion of the equation, see Kevin Rivette and David Klein, *Rembrandts in the Attic* (Boston: Harvard Business School Press, 2000), and Julie Davis and Suzanne Harrison, *Edison in the Boardroom* (New York: John Wiley, 2001). Both books do a nice job of describing some leading practices of companies that extract value from otherwise underutilized IP assets. However, they are silent on how buying IP can help create value for a company. Adam Jaffe and Josh Lerner do an excellent job of summarizing recent research on the history of the U.S. patent office and the policy goals of the U.S. patent system; see *Intellectual Property and Its Discontents* (Princeton, NJ: Princeton University Press, 2004). The authors report that the organization is failing in its task to issue high-quality

patents and to refuse to issue junk patents where no real discovery or value has been created. While the analysis is excellent, the proposed reforms they proffer for the U.S. patent office are surprisingly modest.

12. My previous book is *Open Innovation: The New Imperative for Creating and Profiting from Technology* (Boston: Harvard Business School Press, 2003).

13. Specific sources by these experts include Gordon Petrash, "Dow's Journey to a Knowledge Management Culture," *European Journal of Management* 14, no. 4 (1996): 365–373; Patrick Sullivan, *Value Driven Intellectual Capital: How to Convert Intangible Assets into Market Value* (New York: John Wiley, 2000); and Rivette and Klein, *Rembrandts in the Attic*.

14. One example of such a failed patent-licensing Web site is PLX.com, a company that was profiled in a Harvard Business School case. See Henry Chesbrough and Edward Smith, "The Patent & License Exchange: Enabling a Global IP Marketplace," Case 601-019 (Boston: Harvard Business School, 2000). Another site that has transformed itself is Yet2.com, which is still in operation but has shifted its focus from patent transactions to technology identification and search.

15. Larry Huston and Nabil Sakkab, "Connect and Develop: Inside Procter & Gamble's New Model for Innovation," *Harvard Business Review*, March 2006, 58–67. This lead article provides an in-depth look at P&G's innovation process, with some tantalizing anecdotal evidence of business results. However, there is less evidence of real business benefits (how much sales have increased, how much profits have gone up, the total return on investment for adopting this approach) in this article than one would hope to see.

16. Not shown in figure 1-3 is a key third dimension of innovation economics, the risk of innovation. More open models can address not only cost and time issues, as noted in the P&G article in the previous footnote, but also issues of risk. As we shall see in chapter 6, innovation intermediaries enable companies to treat innovation investment as a variable cost, instead of as a fixed cost. In some models, customers only have to pay when a solution has been delivered. This shifts the risks of failure to others in the system and enables companies to broaden their portfolio of innovation opportunities. Risk management is thus enhanced by Open Innovation, but this point is suppressed in the graph to make the argument about time and cost easier to understand.

17. Rob Hof, "The Power of Us," *BusinessWeek*, June 20, 2005, http://www.innocentive.com/about/media/20050620_BW_FutureTech.pdf.

18. While most observers regard Lemelson's behavior as anti-innovative, the larger question of when an IP owner is advancing or inhibiting innovation is harder to determine. Inventors rely on initial secrecy to establish an advantage in the market. And society grants an inventor a legal right to exclude others (through a patent) for a limited period of time to encourage inventors to take

the risks necessary to achieve a new invention. But society also publishes patents when they are awarded, so that others "practiced in the art" can learn from and build on this invention, perhaps to create further inventions. Publication is key to diffusing new inventions out to the rest of society. Practices such as Lemelson's, which kept the claimed invention secret for decades, deprive society of any publication benefit. So balancing secrecy with diffusion is a challenging social trade-off.

In many situations, though, businesses will have private incentives to publish or diffuse their inventions. These actions could help set a standard or deny a competitor the ability to claim IP for a particular technology. These are explored in more depth in my previous book, *Open Innovation*, particularly chapter 6.

Chapter 2

1. This chapter makes several references to my previous book, *Open Innovation: The New Imperative for Creating and Profiting from Technology* (Boston: Harvard Business School Press, 2003).

2. This data is for the United States. Thomas Friedman's book, *The World Is Flat: A Brief History of the 21st Century* (New York: Farrar, Strauss, and Giroux, 2005), carries this argument through to an international level. Though his data is largely anecdotal, Friedman's claim that "the world is flat" is another way of saying that there are fewer economies of scale in R&D globally as well as in the United States, creating a more level playing field for non-U.S. firms to compete.

3. See Gordon Moore's chapter titled "Some Personal Perspectives on Research in the Semiconductor Industry," in *Engines of Innovation: U.S. Industrial Research at the End of an Era*, eds. Richard S. Rosenbloom and William J. Spencer (Boston: Harvard Business School Press, 1996).

4. A more complete discussion of this transition can be found in *Open Innovation*, ch. 5. See also Emerson Pugh's excellent history of IBM, *Building IBM* (Cambridge, MA: MIT Press, 1995), and Lou Gerstner's informative account of leading IBM through this transition, *Who Says Elephants Can't Dance?* (New York: HarperCollins, 2002).

5. Nabil Sakkab, "Connect & Develop Complements Research & Develop at P&G," *Research-Technology Management* 45, no. 2 (2002).

6. Ibid.

7. Julie Davis and Suzanne Harrison, *Edison in the Boardroom* (New York: John Wiley, 2001), 146.

8. The Tufts Center for Study of Drug Development conducts periodic studies on the costs of new drug development and the attrition rate of compounds in the drug development process. See, for example, Antonio DiMasi,

"Risks in New Drug Development: Approval Success Rates for Investigational Drugs," *Clinical Pharmacology & Therapeutics* 69, no. 5 (2001): 297–307.

9. For an instance of this loose coupling at Microsoft between research and the rest of the organization, see the comments that Rick Rashid, senior VP of research at Microsoft, made at a PC Futures conference in 1998 ("Speech Transcript, Rick Rashid PC Futures 1998," Microsoft Press Room, 1998, https://www.netscum.dk/presspass/exec/rick/6-11pcfutures.mspx). Rashid stated at that event:

> [Microsoft is] really like a university. We're really more like a Stanford University, Carnegie Mellon or MIT in terms of the organizational model. That's the structure we've put together. It's a very open environment. If you go to our web pages, you see all the research that we're doing, see all the names of the researchers that work for us. Nobody reviews people's publications before they're published. We have hundreds of people coming through all the time. At any given point in time we have anywhere between, you know, a dozen to right now we've got about 80 different interns in at the PhD level from all over the world. So it gives you a feeling for what the organization is like. And we've really developed a very strong reputation very quickly, which is, again, very unusual in this sort of basic research academic world.

10. See Rick Rashid's comments, ibid., for an instance of evaluating industrial research primarily through publication metrics. Rashid himself is a respected academic researcher, as that quotation shows.

11. For details on this work, see my article, "Graceful Exits and Foregone Opportunities: Xerox's Management of Its Technology Spinoff Organizations," *Business History Review* 76, no. 4 (2002): 803–838.

12. In two cases I have studied, the interval was quite different. In Lucent's New Ventures Group in the late 1990s, the interval was initially nine months and later condensed to three months, in which the business units had the right of first refusal. In Procter & Gamble, the interval is set at three years after a patent is issued to P&G. If the technology is not in use in at least one P&G business by then, the technology is made available to any outside organization (see Sakkab, "Connect & Develop Complements Research & Develop at P&G"). In both cases, though, the business unit faces competition for using an internal technology or idea and bears some risk if it chooses not to use that technology or idea.

13. One organization that was a poster child for inside-out corporate venture capital is the Lucent New Ventures Group (NVG). This group was profiled in chapter 7 of my previous book, *Open Innovation*. Subsequently, that group was spun off of Lucent—because of Lucent's financial difficulties, the need to finance the portfolio of ventures, and the growing compensation received by the

NVG team, while Lucent was laying off thousands of people. Some conclude from this that the model is deeply flawed. But later events belie that conclusion. The NVG team is now New Venture Partners and sources venture opportunities from three laboratories—Lucent, British Telecom, and Philips—in addition to occasional ventures from other companies' labs. So the model is actually doing very well, albeit in a different ownership and governance model.

14. In "Absorptive Capacity: A New Perspective on Learning and Innovation," *Administrative Science Quarterly* 35, no. 1 (1990), Wesley Cohen and Daniel Levinthal argue that internal R&D increases the "absorptive capacity" of the firm. That is, internal R&D increases the ability of firms to make use of external knowledge in their surrounding environment. Nathan Rosenberg, in his article titled "Why Do Firms Do Research (With Their Own Money)?" (*Research Policy*, 19, no. 7 [1990]: 165–174), has also concluded that firms must have internal R&D to keep up with the research output of others. Historically, companies like IBM and Xerox have initiated internal research and development organizations to help them incorporate advances in the underlying technology base of their industries—a technology insurance policy, if you will, to guard against premature obsolescence.

15. The following description relies heavily on the book by Jerrold Kaplan, *Startup: A Silicon Valley Adventure* (New York: Houghton Mifflin, 1994). While Kaplan is hardly a disinterested observer, the basic facts included here were cross-checked with trade press articles from that period, and Kaplan's assertions accord reasonably well with those articles. However, Kaplan's account does not grapple with a fundamental problem in his business model: his company's success required a key potential competitor (Microsoft) to ally with his business by agreeing to make applications software. This was always destined to be a difficult challenge.

16. Jerrold Kaplan, phone interview by the author, Berkeley, CA, October 17, 2005.

17. Kaplan records that GO's thinking was heavily influenced by the suit initiated by Apple against Microsoft for Microsoft's alleged infringement of the Macintosh look and feel by Microsoft Windows. At the time, each side had spent more than $5 million on the lawsuit, with no resolution of the dispute. Given the high cost, lengthy time, and uncertain outcome of a lawsuit, GO's board decided not to proceed. In my interview with the author (see previous note), Kaplan was not able to discuss the reasons why the company did not pursue litigation at any further length, owing to a pending matter involving Kaplan, GO, and Microsoft's alleged anticompetitive behavior.

18. A virtue of the NDA is that it provides an archival, objective resource to measure how companies protect trade secrets in managing their IP. Since

NDAs constitute a key means for start-up firms to protect against unwanted disclosure, even small start-up companies keep good records of them. These companies often provide documentation of specific conversations with external individuals and firms. Looking back over these NDAs is like seeing a series of snapshots in the growth of a company.

19. This analysis treats all NDAs signed by Collabra as equivalent. As an anonymous reviewer pointed out, this is not the case. One difference is whether the NDA was "one way" (protecting a specified disclosure by one party but not the other) or "two way" (protecting specified disclosures by both parties). Another is the length of the NDA. Collabra's NDA was a single page, while others' NDAs (such as that of IBM) were four pages in length. The longer NDAs were more complex, with more nuanced differentiation regarding what was protected, the duration of the protection, and the scope of protection. A more practical distinction is that many recipients signed Collabra's NDA, while other times Collabra signed the NDA of another firm (such as Microsoft, Novell, or IBM).

20. Collabra had more than eighty-five employees at the time of its acquisition by Netscape in October 1995. The discrepancy between the fourteen NDAs in this dataset and the total number of employees appears to be accounted for by employee agreements signed by all Collabra employees. These employee agreements, however, were not available to me.

21. Information sharing by VCs also allows them to validate their own impressions of the desirability of a potential start-up investment and to discuss the pricing, valuation, and terms of the investment. All of this would be harder if NDA restrictions were placed on them. In addition, offering part of a deal to another investor group increases the likelihood to be invited in return to invest in one of that group's deals. This helps the VC firm develop its "deal flow." While this is undoubtedly good for VCs, the impact on their start-up companies is less clear, since those companies do not participate in the other deals.

22. See Eric Raymond, *The Cathedral and the Bazaar*, 2nd ed. (Sebastapol, CA: O'Reilly, 2001). For a critique of the book, see Nikolai Bezroukov, "A Second Look at the Cathedral and the Bazaar," First Monday, http://firstmonday.org/issues/issue4_12/bezroukov/.

23. Siobhan O'Mahony, "Guarding the Commons: How Community Managed Software Projects Protect Their Work," *Research Policy* 32, no. 7 (2003): 1179–1198; Josh Lerner and Jean Tirole, "The Simple Economics of Open Source," working paper 7600, National Bureau of Economic Research, Cambridge, MA, 2000, 40; and Joel West, "How Open Is Open Enough? Melding Proprietary and Open Source Platform Strategies" *Research Policy* 32, no. 7 (2003): 1259–1285.

24. One useful listing by Joseph Feller can be found at "Bibliography of Research and Analysis," Open Source Resources, http://opensource.ucc.ie/biblio.htm.

25. For more information on this chain of events, see Wikipedia's entry on the SCO lawsuit at http://en.wikipedia.org/wiki/SCO_v._IBM_Linux_lawsuit.

26. See Sun CEO Jonathan Schwarz's own blog on how Sun coexists with open source, http://blogs.sun.com/roller/page/jonathan/20040721.

27. See Carl Shapiro and Hal Varian's excellent book, *Information Rules* (Boston: Harvard Business School Press, 1999), for a more thorough discussion of providing different versions of software with different levels of features at different prices as a strategy to increase profits from software. The book also has excellent discussions of strategies built around sponsored and open standards, and network effects in the adoption of new technologies.

28. See Lawrence Lessig, *The Future of Ideas* (New York: Vintage, 2002), for a thoughtful and impassioned analysis of the benefits of a creative commons. While Lessig's analysis is trenchant from society's point of view, he underplays greatly the role of firm strategies in the formation of these commons. While the commons may be a good thing for society, companies create them to advance their own goals, which may or may not be in the best interests of society (particularly if companies co-opt the commons to focus solely on their own goals).

29. For a recent and thoughtful treatment of the user side of open source technologies, see Eric von Hippel's *Democratizing Innovation* (Cambridge, MA: MIT Press, 2005). One issue that separates von Hippel's work from this chapter is the question of a *business model*. The term is not found in his book, and he believes that users are often better served by freely revealing their innovations to others in their communities. That can be a precursor to a business model, if the revealing party has complementary activities (where he or she has more protection) that benefit from more innovation. But even in open source, business models are developing and are important to the further adoption of the technologies in society.

30. This will be discussed further in chapter 8. Joel Cawley, VP of corporate strategy for IBM, quantifies the savings like this: "I have long observed that it takes $500 million to create and sustain a commercially viable OS [operating system]. Take our development lab for Linux in Beaverton, Oregon. We spend about $100 million there each year. About $50 million of that is spent on basic improvements to Linux, how to make it more reliable. The other $50 million is spent on things that IBM needs, like special drivers for particular hardware or software to connect with it. We asked the Open Source Development Lab to estimate how much other commercial development spending was being done on Linux. This didn't count any university or individual work, just other companies

like us. They told us the number was $800 [million] to $900 million a year, and that the mix of basic/specific needs was close to 50/50. So that $500 million expense for a viable OS is there now for Linux as well. And we only pay $100 million toward that. So you can see even from a very narrow accounting view that this is a good business investment for us." (Joel Cawley, interview by author, Armonk, NY, October 7, 2005.)

31. For more on open source software's business issues, see my colleague Joel West's article, "How Open Is Open Enough? Melding Proprietary and Open Source Platform Strategies," *Research Policy* 32, no. 7 (2003): 1259–1285.

Chapter 3

1. This section borrows liberally from other research on the history of the U.S. patent system. See the Web site, "US Patent System," Great Idea Finder, http://www.ideafinder.com/history/inventions/story096.htm, for a brief overview. For a more detailed history from 1790 to 1900, see Kenneth Dobyns, *A History of the Early Patent Office*, reprint ed. (Washington, DC: Sergeant Kirkland's, 1997).

2. U.S. Constitution, article 1, section 8.

3. See Adam Jaffe and Josh Lerner's insightful book, *Innovation and Its Discontents* (Princeton, NJ: Princeton University Press, 2004), for an extended discussion of these patent protection issues. On pages 98–101, the authors report evidence that the likelihood of winning a patent protection lawsuit was more than 30 percent overall in the period of 1953 through 1977 but varied widely from one federal district to another. This high variance (from as low as 8 percent in the Great Plains to as high as 57 percent in the Rocky Mountains) has no obvious explanation and strongly encouraged litigants to shop for the best district for their own interests. As Jaffe and Lerner show, Congress gradually became concerned about this state of affairs and set up a court of appeals for the federal circuit in 1982 to centralize patent cases, promote greater specialization of legal knowledge about patents, and eliminate the gross disparities between circuits.

4. Ibid., 106. As Jaffe and Lerner discuss there, comparing the 68 percent rate with the earlier rate is more complicated than looking at the two numbers. It is quite likely that the higher rate of upholding patents by the new court caused people holding weaker patents (who might have avoided litigation entirely under the old regime, since they had a lower chance of winning) to now take their chances in court. This "mix shift" means that more and weaker patents were being litigated than in the 1953–1977 period. Thus the true difference between the old and new patent regimes likely is even greater than the percentages suggest.

5. For more on the history of Kilby and the checkered history of his patent, see Fiscal Notes, Texas Comptroller of Public Accounts, http://www .window.state.tx.us/comptrol/fnotes/fn9810/fn.html, and T. R. Reid, *The Chip: How Two Americans Invented the Microchip and Launched a Revolution* (New York: Simon and Schuster, 1984).

6. For an excellent discussion of TI's experience with licensing, see David Teece and Peter Grindley, "Managing Intellectual Capital: Licensing and Cross-Licensing in Semiconductors and Electronics," *California Management Review* 39, no. 2 (1997).

7. Ralph Gomory, interview by author, Yorktown Heights, NY, October 7, 2005. Gomory is now the head of the Alfred P. Sloan Foundation in New York City.

8. IBM has received the most U.S. patents of any company in the world for eight straight years from 1998 through 2005. However, IBM was not able to do this by simply filing many more patent applications. Its success started from its longstanding commitment to R&D, which constituted more than six thousand people across laboratories on five continents, with spending in excess of $5 billion as of 2004. Companies that benchmark IBM's success with its patents, and seek to emulate it, sometimes neglect the preconditions that supported its success. You cannot become a patent leader overnight. It requires years of recruiting and building talent, transferring knowledge to the market via products and services. Only then can one benefit from a systematic legal effort to file for patents.

9. Paul Horn, senior VP of research and head of IBM's research division, interview by author, Yorktown Heights, NY, October 7, 2005.

10. Joel Cawley, VP of corporate strategy at IBM, interview by author, Armonk, NY, October 7, 2005.

11. See Ashish Arora, Andrea Fosfuri, and Alfonso Gambardella, *Markets for Technology* (Cambridge, MA: MIT Press, 2001). The bulk of the authors' research considers the development of markets for technology in the chemicals and petrochemicals industry. But their more general point is that there are many situations where economic markets have developed in which a supplier of a technology can sell that product to others that turn around and use it to develop new products. A related model can be found in Joshua Gans, David Hsu, and Scott Stern, "When Does Start-Up Innovation Spur the Gale of Creative Destruction?," *RAND Journal of Economics* 33, no. 6 (2002): 571–586. That article analyzes when an inventor of a new technology should use it to compete in a product market as a new competitor and when the inventor should instead choose to sell it to the existing competitors in the market.

A more recent article by Ashish Arora and Robert Merges, "Specialized Supply Firms, Property Rights and Firm Boundaries," *Industrial and Corporate Change* 13, no. 3 (2004): 451–475, explores the role of new entrant start-up firms in stimulating innovation and discusses conditions under which strong intellectual property rights help enable the formation and entry of such firms. This entry stimulates the growth of intermediate markets for innovation.

12. See my report with David Teece to the Alfred P. Sloan Foundation, "The Globalization of R&D in the Chinese Semiconductor Industry," December 2, 2005. The full report can be found at http://web.mit.edu/ipc/sloan05/.

13. This section is inspired by Michael Jacobides's dissertation research on the mortgage banking industry, which was started while he was at Wharton. He is now a professor at London Business School and has published numerous papers on this topic. One particular paper of interest for this section is "Industry Change Through Vertical Disintegration: How and Why Markets Emerged in Mortgage Banking," *Academy of Management Journal* 48, no. 3 (2005): 465–498.

14. For more information on this analysis of patent reassignments, see the work from me and Alberto Diminin, "Technology Sourcing and the Management of Intellectual Property Rights: Evidence from Patent Reassignments," working paper, Center for Open Innovation, Institute of Management of Innovation and Organization, Haas School of Business, UC Berkeley, 2005.

15. The Dialog database allowed us to query attributes of the reassignments more easily than the USPTO data did directly. We corroborated the accuracy of the Dialog data by spot-checking dozens of reassignments on Dialog with the USPTO data. These were all done correctly, giving us confidence in using the Dialog data. A further benefit of the Dialog data source was that it was able to omit the initial reassignment of a patent from an individual within a company to the company itself. That initial assignment was not of interest to us, and Dialog enabled us to exclude it from the analysis that follows.

16. John Wolpert used to be in charge of IBM's Extreme Blue organization. Currently he is the head of the BRIDGE program, which is part of the InnovationXchange in Australia. This organization will be discussed further in chapter 6.

17. Adrienne Crowther, phone interview with author, March 10, 2005.

18. Eric von Hippel's classic work on this subject is *The Sources of Innovation* (New York: Oxford University Press, 1988). His more recent book, *Democratizing Innovation* (Cambridge, MA: MIT Press, 2005), has deepened his understanding of user-driven innovation processes, with concepts of lead users, user toolkits, and the role of users in open source software communities.

19. A simple example of such a combination that increases efficiency and productivity in storage would be where the customer ran a number of servers and used them about 40 percent. IBM might combine the servers into a pool with other servers and aggregate the customer's storage demands with other customers. It might be able to achieve a 70 percent or 80 percent utilization of its servers, saving customers money while lowering its fixed costs as well.

20. Lemley studied the rules and bylaws of forty-three different standards-setting organizations. He found that there were few "standards" on how these rules and bylaws were developed and that organizations differed from each other in what information disclosure they required among participating companies. In his judgment, these differences in rules were "largely accidental" and not due to any conscious intention on the part of those organizing that standards-setting organization. See Mark Lemley, "Intellectual Property Rights and Standards Setting Organizations," *California Law Review*, December 2002.

21. One could write entire articles and books on the role of standards in technology strategy and competition. Happily, others have done that very thing. Carl Shapiro and Hal Varian's *Information Rules* (Boston: Harvard Business School Press, 1999) is one such book. The research of Michael Katz and Carl Shapiro, and Carl Shapiro and Joseph Farrell, are other seminal sources for such analysis. Mark Lemley has done important work on the legal side of standards, as has Robert Merges. All of the authors listed here are at UC Berkeley, making Berkeley something of an industrial cluster for this kind of analysis.

22. See Fiona Murray's work, "Innovation as Co-evolution of Scientific and Technological Networks: Exploring Tissue Engineering," *Research Policy* 31, no. 31 (2002): 1389–1403. Murray's TABs play an important role in the development and commercialization of science, particularly in the domain of tissue engineering. She shows how the presence of top-rated scientists on TABs is associated with the introduction of more advanced products into the market.

23. Kevin Rivette and David Klein, *Rembrandts in the Attic* (Boston: Harvard Business School Press, 2000).

24. See Deepak Somaya's article "Strategic Determinants of Decisions Not to Settle Patent Litigation," *Strategic Management Journal* 224, no. 1 (2003): 17–38. Somaya used court records of IP disputes to make inferences about whether and when companies would settle with each other or take a dispute through a court trial.

25. Interview with Jeff Weedman at UC Berkeley, April 25, 2005.

26. This section is inspired in part by an article by Michael Kayat and Tova Greenberg, "IP-Based Open Innovation Pre-empts Trolls," *Intellectual Asset Management*, February/March 2006, 43–46. (*Intellectual Asset Management* is online at http://www.iam-magazine.com.) Kayat and I delivered a talk on this subject to the Licensing Executives Society in New York City on September 14, 2006.

Chapter 4

1. Carl Shapiro and Hal Varian's book, *Information Rules* (Boston: Harvard Business School Press, 1999), and the references in that book, are good places to start for the economic and legal aspects of protecting one's IP. Kevin Rivette and David Klein's book, *Rembrandts in the Attic* (Boston: Harvard Business

School Press, 2000), provides an exuberant account of some of the management opportunities in underutilized IP. Another, more balanced account of the management of IP can be found in Julie Davis and Suzanne Harrison's book, *Edison in the Boardroom* (New York: John Wiley, 2001).

2. This analysis is offered to expose managers who ordinarily do not deal with IP to the issues and opportunities that IP offers. Readers should be aware, however, that this analysis is a simplification of reality. For example, the following analysis does not consider the complications created by the fact that most U.S. patents have multiple claims, not just one. Infringing on a single claim of a valid patent is sufficient to be legal infringement. In all of this, any important decisions should be taken only after consulting with legal counsel because the many such details that I am glossing over here may prove critical in a litigation action.

3. Economists distinguish between "simple" and "complex" technologies in terms of whether owning a patent gives you effective protection of your technology. *Simple technologies* are ones like those in biotechnology, where your patent on a particular compound not only excludes anyone else from using that compound but also effectively lets you be the only one to develop it. *Complex technologies* like semiconductors, by contrast, involve hundreds or even thousands of patents held by dozens of companies. No one patent in this sector enables a company to practice the technology, unless that company embarks on extensive cross-licensing activities. See B. Douthwaite, D. H. Keatinge, and R. Park, "Why Promising Technologies Fail: The Neglected Role of User Innovation During Adoption," *Research Policy* 30, no. 5 (2001): 819–836.

4. See Bronwyn Hall and Rosemarie Ham Ziedonis, "The Patent Paradox Revisited," *RAND Journal of Economics* 32, no. 1 (2001): 101–128. In that paper, the authors discuss their explanation for the rise in semiconductor patenting as a result of the strengthened patent courts, which in turn strengthened the need for more patents for defensive purposes.

5. Michael Kayat and Tova Greenberg credit Peter Detkin, formerly of Intel and now at Intellectual Ventures, for coining the term *troll*. See their article, "IP-Based Open Innovation Pre-empts Trolls," *Intellectual Asset Management*, February/March 2006, 43.

6. As an anonymous reviewer pointed out, patent maps can be misleading because the software that generates them usually does not take the many individual claims within each patent into account. Mapping all of the claims is a daunting task and so is typically not done. In a legal action, however, it is the individual claims that will control a dispute, so the mapping may gloss over liabilities that could surface later on in court. Still, mapping patents is surely better than doing nothing, even if important risks are left out of the analysis.

7. See William Abernathy and James Utterback, "Patterns of Industrial Innovation," *Technology Review*, June/July 1978, for the pioneering work in this

area. These authors show that technological competition goes through a predictable cycle. First, very different technologies compete in the market until a dominant design emerges within the market. Then there is a marked phase shift, and competition moves within the dominant design, as those designs that did not win fade from the market. At that point, the basic parameters of the technology have been established and companies must compete on process technologies to achieve success. A later book by Utterback, *Mastering the Dynamics of Innovation* (Boston: Harvard Business School Press, 1994), documented this cyclical pattern in a wide range of industries, from sailing ships to ice and refrigeration. While the losing technologies often demonstrated a "last gasp" of improvement, established players that failed to support the newly dominant design went out of business.

Important contributions subsequently have been made by Philip Anderson and Michael Tushman, in "Technological Discontinuities and Dominant Designs: A Cyclical Model of Technological Change," *Administrative Science Quarterly* 35, no. 3 (1990), who showed that some dominant designs were "competency enhancing" and caused incumbent firms to become further entrenched in their industry. Other designs were "competency destroying," which caused incumbent firms to be dislodged in the market. Steven Klepper, in "Entry, Exit, Growth and Innovation over the Product Life Cycle," *American Economic Review*, 87 no. 7 (1996): 562–583, documented how firm size was initially disadvantageous in early stages of the life cycle, but strongly advantageous in later stages. In these later stages, the ability of large firms to spread costs over more volume was a powerful advantage in the market.

These are largely supply-side views. A demand-side view of the technology cycle comes from patterns of technology diffusion (Everett Rogers, *Diffusion of Innovations* [New York: Simon and Schuster, 1995]), where early adopters are followed by an early majority, then a late majority, and finally the laggards. Ron Adner and Daniel Levinthal, in "Demand Heterogeneity and Technology Evolution," *Management Science*, 47 no. 9 (2001): 611–628, propose a demand-side model that explains pre- and postdominant design cycles from the consumer's perspective. In their model, consumers are unsatisfied with the performance of the technology in the early stage of the cycle. Gradually, as suppliers advance the technology, consumers become more satisfied, until a threshold of sufficiency is reached. Past that point (which shifts the cycle), further performance improvement is valued little, if at all, by consumers.

Whatever the sources of the technology cycles, their presence is well accepted in academic and business circles. What has not been developed until now is the implications for this cycle on the management of intellectual property.

8. See Donald Sull's fascinating research, "The Dynamics of Standing Still: Firestone Tire & Rubber and the Radial Revolution," *Business History Review*

73, no. 4 (1999): 430–464, for an engaging yet scholarly account of how U.S. tire manufacturers responded to the threat posed by the new and superior radial tire technology offered by Michelin and others. Sull's research finds that while the U.S. manufacturers tried many maneuvers to fend off the threat, they failed to make the critical and necessary investments to adopt the technology until very, very late in the game.

9. See David Teece, "Profiting from Innovation," *Research Policy* 15, no. 6 (1986), for a seminal treatment of how first movers do or do not profit from their innovations. In Teece's analysis, it is not whether the firm is the first to offer the technology that matters. Instead, he shows the importance of complementary assets (such as manufacturing, marketing, or distribution assets) in whether first movers in a new technology succeed in capturing value from their innovations, or whether "fast followers" overtake them and capture the lion's share of the profits instead.

10. Apple's success with its iPod has allowed it to create more complementary assets, in the form of Apple retail storefronts that sell the iPod, along with other Apple products. This growing retail presence will become a stronger complementary asset for Apple in future technologies it may choose to offer.

11. As David Teece noted in "Profiting from Innovation," IP protection can come from business assets and business models, as well as the courts. In this case, Apple's strong brand, marketing, distribution, and operations assets all help capture value from the iPod and to date have withstood efforts by competitors to imitate it. Interestingly, though, Microsoft actually filed a patent on an iPod-type device before Apple did. How that potential patent infringement situation will be resolved remains to be seen as of this writing.

12. James Utterback's work, particularly his book, *Mastering the Dynamics of Innovation*, contains many examples of the futility of trying to hang on to the losing technology.

13. If the firm has other value capture points in its business model (such as a strong brand, world-class manufacturing, or an excellent sales force), then it may not need its IP managed so as to capture value.

14. The rise and fall of IBM in the PC business is well told in Charles Ferguson and Charles Morris's book *Computer Wars* (New York: Times Books, 1993). As the authors point out, IBM never even owned the source code for the DOS operating system that it purchased from Microsoft. In "When Is Virtual Virtuous?" *Harvard Business Review*, January–February 1996, David Teece and I analyzed the benefits and risks of IBM's outsourcing a systemic, interdependent technology like the operating system. Robert Purvy, a former Xerox employee who now works as a patent attorney, offered this assessment of IBM's situation in the PC market: "IBM had the misfortunate to develop their stuff *barely* be-

fore software patents became acceptable (1982 or so). Nowadays, they'd patent the very *idea* of a ROM BIOS and all its functions!" (author's phone interview with Robert Purvy, January 21, 2005). His point is that the development of the PC market might have proceeded differently in such a world.

15. This ability to extract some value from exiting a business also suggests that companies should be careful how strongly they criticize the "trolls" who profit from owning patents without practicing them. In the cases noted earlier in this chapter, GE and IBM acted like these trolls when they exited their businesses. Presumably, they would like to retain the ability to receive compensation from other IP if and when they choose to exit future businesses.

16. See en.wikipedia.org/wiki/Microsoft.

17. Microsoft cofounder and chairman Bill Gates made the following observation in an unguarded moment in 1998 to an audience at the University of Washington: "Although about 3 million computers get sold every year in China, people don't pay for the software. Someday they will, though. And as long as they're going to steal it, we want them to steal ours. They'll get sort of addicted, and then we'll somehow figure out how to collect sometime in the next decade." (Charles Pillar, "How Piracy Opens Doors for Windows," *Los Angeles Times*, April 9, 2006, C1.) The account I have provided in this chapter adds the importance of network effects to become the desktop standard and the need to win the battle against Linux on the desktop first, though these factors are certainly more important now than in 1998.

Chapter 5

1. For most companies, it can be said that the company truly has a single business model. Even companies the size of Intel, with different businesses, are driven largely by the same business model across these businesses. There are a few companies, though, that join together different businesses (and different models) in a conglomerate. General Electric and Berkshire Hathaway are two of the best known examples of these. In these latter instances, it may be more helpful to analyze the business models of individual business units, rather than the whole company. At the corporate level, there is a separate analysis needed for how to manage different business models within a single corporation. I am grateful to an anonymous reviewer for stimulating me to clarify this point.

2. For a discussion of business models in the context of the Internet, see Allan Afuah and Christopher Tucci, *Internet Business Models* (New York: McGraw-Hill/Irwin, 2000), and Raffi Amit and Christophe Zott, "Value Creation in e-Business," *Strategic Management Journal* 22, no. 6 (2001). Both of these sources define a *business model* as how a company intends to make money in

the hothouse of the many Internet technologies available. Richard Rosenbloom and I, in "The Role of the Business Model in Capturing Value from Innovation," *Industrial and Corporate Change* 11, no. 3 (2002), discuss business models in the context of Xerox and its storied history in copiers and printers, and its less successful history in computers. While we agree that business models involve an intention to convert technology into money, we point out the logical antecedents of such an intention (such as the selection of the target market and the value proposition offered to that market), as well as the means to reach the market (such as the value chain and surrounding value network). More recently, Joel West and Scott Gallagher's work, "Challenges of Open Innovation: The Paradox of Firm Investment in Open Source Software," *R&D Management* 20, no. 5 (2006): 315–328, has shown a variety of business models in open source software, where companies using open source software are constructing businesses to make money from surrounding activities.

3. According to the National Science Foundation (NSF), 16.7 million domestic employees worked in U.S. companies of five or more people conducting R&D. (See Table A-34. Domestic employment of companies that performed industrial R&D in the U.S., by industry, by size of company: 2001, http://www .nsf.gov/statistics/nsf05305/tables/taba34.xls.) A different NSF analysis tracked 14.5 million people working in the United States in companies spending $1,000 or more on R&D. (See "US Corporate R&D Investments, 1994-2000, Final Estimates," http://www.technology.gov/reports/CorpR&D_Inv/CorpR&D_Inv1994-2000.pdf.) By comparison, more than 130 million people were in the U.S. workforce in 2000. (See "Total U.S. Employment by Industry 1990-2000," http://www.bizstats.com/employment.htm.) Note that individual inventors are not included in this data, nor are government, education, or university employees.

4. See the article in *Strategy and Leadership* by Lotfi Belkhir, Liisa Valakangas, and Paul Merlyn on innovation and companies that have only one successful product. "One CEO's product development motto: care for innovations like newborns," http://www.kirtas-tech.com/uploads/other/StrategyLeadership.pdf. The authors identify Corio and Palm Computing as two examples of one-hit companies. Steve Ballmer has stated that Google might also become a one-hit company. See "Is Google a One Hit Wonder?" *The Motley Fool*, May 23, 2005, http://www.fool.com/News/mft/2005/mft05052301.htm, for coverage of his remarks at Stanford University. The business model framework provides one means to assess a company's ability to innovate beyond its initial product. While I have not studied Google personally with this model, many of my students have. Their assessment to date is that Google is far beyond type 2 and thus likely to sustain itself well beyond the one-hit company scenario.

5. This is a fun parlor game for a rainy night with friends. See http://www.onehitwondercentral.com for listings of such songs by decade, starting from the 1950s.

6. As discussed in chapter 2, some of this internal R&D activity, however, does not fit the business model and is left on the shelf. This waste is viewed as a cost of doing business in this stage of the BMF.

7. I discussed this situation at length in chapter 1 of *Open Innovation: The New Imperative for Creating and Profiting from Technology* (Boston: Harvard Business School Press, 2003). A more detailed historical account may be found in my article "Graceful Exits and Foregone Opportunities: Xerox's Management of Its Technology Spinoff Organizations," *Business History Review* 76, no. 4 (2002): 803–838.

8. These three benefits—saving cost and time, and reducing risk—may sound too good to be true. It certainly will take time, skill, and some experimentation to realize these benefits from a more open business model. But there is growing evidence that open business models can in fact have these attributes. Open business models help the firm accelerate its time to market by leveraging external projects that enhance the internal road map, and that have often been under way for months or even years. This gives the project using those external resources a head start and lower development costs, since those costs are shared with other parties. In chapter 8, we will discuss IBM's efforts in open source software and how that company is obtaining a $500 million operating system per year for only $100 million per year. That is a tremendous cost savings. And leaders like Dell and Wal-Mart share supply-chain risks with their suppliers.

9. Kraft Foods, for example, has created a new leadership position called the "senior vice president for Open Innovation." Mary Kay Haben, a twenty-five-year veteran of Kraft, is the first senior vice president of Open Innovation and has made it clear that Kraft is eager to solicit marketing and development opportunities with outside partners. See Betsy Spethmann, "Kraft Goes Experiential," *Promo*, January 18, 2006, http://promomagazine.com/news/kraft_innovation_011806/.

10. IBM created a venture capital operation to promote IBM to start-ups as a partner of choice. Instead of investing capital in young start-ups, IBM tried to convince them to use its products and services.

11. See David Teece, Gary Pisano, and Amy Shuen, "Dynamic Capabilities and Strategic Management," *Strategic Management Journal* 18, no. 7 (1997): 509–533, for the seminal article on dynamic capabilities.

12. For two excellent books on the topic of platforms and building the surrounding ecosystem, see Annabelle Gawer and Michael Cusumano, *Platform*

Leadership (Boston: Harvard Business School Press, 2002), and Marco Iansiti and Roy Levien, *The Keystone Advantage* (Boston: Harvard Business School Press, 2004).

Chapter 6

1. In my view, the far more likely outcome is that the potential idea or technology would have simply remained unused, and the company would have received no value from it whatsoever. Remember the very low utilization rates of internal patented technologies that were discussed in chapters 1 and 2.

2. Bill Breen, "Lilly's R&D Prescription," *Fast Company*, April 2002.

3. Author's interview with Jill Panetta, InnoCentive's corporate offices, Waltham, MA, on June 11, 2003.

4. This value was a conservative estimate, based on eight of the twelve challenges. The other four were estimated to yield a net benefit of zero, due to difficulties in accurately attributing costs and revenues to these latter projects. If some positive value were assigned to those as well, the cost-benefit ratio would have been even higher. However, it ignores other work done by Lilly after receipt of the solution. This work also contributed significantly to the $8.8 million.

5. It may be that InnoCentive's clients need to make some organizational changes in order to receive the full benefit of these search services. This may be an answer to this puzzle. See note 13.

6. Paul Kaihla, "Building a Better R&D Mousetrap," *Business 2.0*, September 2003, http://www.business2.com/articles/mag/current/0,1639,,00.html. Conversations with the company indicate that this network of solvers expanded to more than eighty thousand participants as of mid-2005.

7. While the terms of these agreements varied in their details, the general provisions gave academic scientists in each country the right to participate as solvers in InnoCentive's process and gave seekers full rights to the solutions proposed. In turn, the academic institution would receive a portion of the award payment, while the solver would receive the rest of the payment.

8. Author's interview with Ali Hussein, InnoCentive's corporate offices, Waltham, MA, on June 11, 2003.

9. Mohan Babu, "In the Global Web from Home Ground," http://www.deccanherald.com/deccanherald/july28/eb5.asp.

10. NineSigma homepage, http://www.ninesigma.com.

11. Author's interview with Mehran Mehregany, Philadelphia, PA, May 14, 2004. Professor Mehregany also holds the Goodrich Professor of Engineering Innovation chair at Case Western Reserve University School of Engineering.

12. John Teresko, "Open Innovation? Rewards and Challenges," *Industry Week*, June 1, 2004, http://www.industryweek.com/CurrentArticles/asp/articles.asp?ArticleId=1627.

13. Ibid. This insight, that organizational change is required before the real benefits of external technologies can be realized, may help explain the challenge of scaling up InnoCentive's offerings. InnoCentive has delivered many solutions to its clients and has some tangible evidence that the value of these solutions greatly exceeds their cost, yet the company is having difficulty getting clients to use its service more. InnoCentive may need to add an organizational change component to its service offering to achieve greater volumes of business from its clients.

14. The information in this section regarding BIG comes from three primary sources. First, a wonderful case study, "What's the BIG Idea? (A)," was written about the company by Clay Christensen and Scott Anthony (Case 9-602-105 [Boston: Harvard Business School, 2001]). Second, I taught this case on three different occasions while I was at Harvard Business School and was fortunate enough to have both Michael Collins, the founder of BIG, and Alex D'Arbeloff, the chairman of BIG, come to class on two of those occasions. Their discussion with me before and after class was also very helpful. Finally, the BIG Web site (http://www.bigideagroup.net) was useful for updated information.

15. Christensen and Anthony, "What's the BIG Idea? (A)," 4.

16. Ibid., 5–6.

17. This high renewal rate is perhaps the most encouraging measure of the model's performance to date. The high rate is even more impressive when one considers that the initial membership fee was subsidized in order to recruit some small companies, as well as large ones. The renewals, though, received no subsidy whatsoever. This suggests that the IXC model is seen as economically justifiable at full market rates. However, in 2006, Wolpert left IXC to pursue a PhD.

18. Author's interview with Keith Cardoza, Ocean Tomo corporate offices, Chicago, IL, October 26, 2005. For those of us not expert in finance, this is what *alpha* means: "Alpha is a risk-adjusted measure of the so-called 'excess return' on an investment. It is a common measure of assessing an active manager's performance. The difference between the fair and actually expected rates of return on a stock is called the stock's alpha." [Source: "Alpha," Wikipedia, http://en.wikipedia.org/wiki/Alpha_(Investment).]

19. This is entirely consistent with the data presented in chapter 3 on patent reassignments. There, we saw that the fourth most frequent reason for selling semiconductor patents was to obtain a security interest. Here, we see a firm that specializes in helping borrowers work with lenders where IP is to be part of the collateral in the financing provided.

20. Author's interview with Jim Malakowski, Ocean Tomo corporate offices, Chicago, IL, October 26, 2005.

Chapter 7

1. Most of the research in this book has been done by me personally, through interviews, discussions, and often multiple interactions with the organizations. With Qualcomm, however, a history by David Mock has emerged called *The Qualcomm Equation* (New York: Amacom, 2005). Mock's history of this company is impressive in its detail and thoroughness. He enjoyed tremendous access to the founders and key managers of the company and did a very good job of documenting what he learned. It seemed pointless to try to recreate his efforts. In true Open Innovation fashion, I have chosen instead to try to build on them and explore the business model of Qualcomm more precisely.

2. By comparison, Intel—a very profitable company—had about $425,000 of revenue per employee for 2005. Broadcom, a successful semiconductor firm that outsources its manufacturing, had about $580,000 of revenue per employee in 2005.

3. In chapter 2, we examined the low rate of utilization of patented technologies inside companies. Here, a similarly low utilization rate is observed for invention disclosures at universities. Note, though, that invention disclosures must be patented *before* their being licensed, so comparing those measures is mixing apples and oranges. A more directly comparable statistic to the utilization measure in chapter 2 would be to find out what percentage of invention disclosures are patented and then what percentage of patented inventions are subsequently licensed. At UC Berkeley, for example, the university policy is to have one or more corporate partners defray the costs of patenting, so few of the patented inventions go unlicensed, since the corporation would not pay for the patenting costs unless it intended to license the technology subsequently.

4. Quotation from Clifford Gross is from a phone interview with then-doctoral student Alberto Diminin on February 21, 2005; Clifford Gross, Uwe Reischl, and Paul Abercrombie, *The New Idea Factory* (Columbus, OH: Battelle, 2000).

5. According to the company's 2004 Form 10-K: "[Of] the total capital invested in our intellectual property acquisition companies during the year ended December 31, 2004, $50,000 was expended on research and development costs, $753,700 was expended on license and consulting fees, and $347,500 remained in our intellectual property acquisition companies at the time they were sold to our portfolio companies. All of these items are reflected in the Consolidated Statement of Operations as sales and marketing expenses." (Source: UTEK, Form 10-K, 2004, 21.)

6. Intellectual Ventures homepage, http://www.intellectualventures.com.

7. This quotation from Greg Gorder, and the ones that follow, are from author's interview, Intellectual Venture headquarters, Redmond, WA, May 10, 2005.

8. Author's interview with Laurie Yoler, law offices of Heller, Ehrman, in Palo Alto, CA, June 7, 2005. Yoler has since left the company.

9. Michael Kanellos, "Patent Firm Woos Big-Name Inventors," CNET, April 21, 2006, http://news.com.com/Patent+firm+woos+big-name+inventors/2100-11746_3-6063457.html.

10. This quotation from Nathan Myhrvold and subsequent ones—from him and Clarence Tegreene—are from conversations with the author after the invention session, Intellectual Ventures Headquarters, Redmond, WA, May 10, 2005.

11. This raises the question of whether Intellectual Ventures has actually met the patent office's requirements of an invention "reduced to practice" for the inventions it has claimed. Ultimately, this is an empirical question that will be decided by the U.S. Patent and Trademark Office on each patent application. The company has already received four patents but has hundreds more in process that have not yet been issued. The question may be tested again in court, if and when Intellectual Ventures attempts to enforce its patent rights against recalcitrant infringing companies. On June 26, the company announced its five hundredth patent application.

12. In June, 2006, Microsoft announced a $36 million equity investment in IV and further agreed to license up to $80 million of IV's inventions. It is likely that other investors will identify themselves in the future.

13. This definition is taken from my previous book, *Open Innovation: The New Imperative for Creating and Profiting from Technology* (Boston: Harvard Business School Press, 2003), 64–65. See the article Richard Rosenbloom and I wrote, "The Role of the Business Model in Capturing Value from Innovation," *Industrial and Corporate Change* 11, no. 3 (2002), for a more academic treatment of this definition, other academic approaches to the concept, and its roots in earlier business strategy literature. More recent work on business models can be found in Raffi Amit and Christophe Zott, "Value Creation in e-Business," *Strategic Management Journal* 22, no. 6 (2001): 493–520, and Michael Rappa, "The Utility Business Model and the Future of Computing Services," *IBM Systems Journal* 41, no. 1 (2004): 32–42. For an academic survey of the topic, see Jonas Hedman and Thomas Kallings, "The Business Model Concept: Theoretical Underpinnings and Empirical Examples," *European Journal of Information Systems* 12, no. 1 (2003): 49–59.

Chapter 8

1. See John Burgess, "IBM's $5 billion loss highest in US Corporate History," January 20, 1993, http://www-tech.mit.edu/V112/N66/ibm.66w.html.

2. Gerstner's own account of his years at IBM can be found in Lou Gerstner, *Who Says Elephants Can't Dance?* (New York: HarperCollins, 2002).

Inevitably, perhaps, this account centers on Gerstner's own actions and slights the many contributions of others in the IBM organization that propelled IBM forward. Nonetheless, a wide variety of IBM executives credit Gerstner and the executives that he brought in from outside as being instrumental in developing a more open business model. See the quotations that follow in this section. Other accounts of IBM's transformation can be found in chapter 5 of my previous book, *Open Innovation: The New Imperative for Creating and Profiting from Technology* (Boston: Harvard Business School Press, 2003), and in Robert Austin and Richard Nolan's case study of IBM, "IBM Corporate Turnaround," Case 600-098 (Boston: Harvard Business School, 2000).

3. Author's interview with Joel Cawley, VP of corporate strategy at IBM, Armonk, NY, October 7, 2005. Cawley observed, "[Our need for revenue] prompted us to find other partners who also needed leading-edge semiconductor technology. So today we license AMD, Cisco, [and] Sony, and net their patent portfolio against ours, with a balancing payment. Overall, we break even on this or make a little money. It costs us $500 million to get it, and we get more than $500 million more from others. But this was not the result of any grand strategy at a corporate level. It is the result of a business responding to the needs of its industry and its own need to stay profitable."

4. Author's interview with Jerry Rosenthal, IBM's Yorktown Heights, NY, Research Laboratory, October 7, 2005.

5. Comments made by Catherine Lasser, Joel Cawley, Jerry Stallings, Paul Horn, and Jesse Abzug are from interviews by author, IBM's Yorktown Heights, NY, Research Laboratory (Horn, Lasser, Abzug, and Stallings) and IBM Global Headquarters (Cawley), October 7, 2005.

6. For an account of Gerstner's address to the eBusiness Conference and Expo in New York, see Joe Wilcox, "IBM to Spend $1 billion on Linux in 2001," December 12, 2000. http://news.com.com/2100-1001-249750.html.

7. This doesn't mean that IBM has stopped trying to earn money on its IP. IBM's strategy combines the pursuit of revenue and profit from its IP, along with strategic donations of other IP. IBM now thinks of its IP more strategically. In some cases, maximizing the financial value of IP might close off opportunities to gain a stronger position in a new market or ecosystem, according to IBM's Paul Horn. The open source initiative illustrates this. IBM gives away development tools to lower the cost of developing for Linux and recruits more people into that initiative. At the same time, IBM develops many proprietary technologies, such as middleware, that benefit directly from more open source software. Giving away one helps increase sales of the other—an enlightened but nonetheless very profit-oriented business model.

8. Kevin Rivette is the same person who coauthored a book on IP management in 2000 called *Rembrandts in the Attic* (Boston: Harvard Business School Press).

9. Comments made by Gil Cloyd, Steve Baggett, Jeff Henner, Jeff Weedman, Martha Depenbrock, and Mike Hock are from interviews by author, P&G Headquarters, Cincinnati, OH, on October 20, 2005.

10. Larry Huston's remarks were made in a talk he delivered to the Mack Technology Center at The Wharton School at the University of Pennsylvania, May 14, 2004.

11. In addition to the interviews, additional information on P&G's Reliability Engineering comes from Suzanne Harrison and Patrick Sullivan's book, *Einstein in the Boardroom* (New York: John Wiley, 2006), 126.

12. Through the 1980s, Glad was owned by Union Carbide. In the 1990s, though, the brand was spun off into a consumer products firm, First Brands. First Brands was acquired by Clorox in 1999. At the same time, the Ziploc and Saran Wrap brands were sold by Dow Chemical to SC Johnson. So there was something of a pattern of chemical companies selling off consumer brands to consumer products companies. In 2006, P&G was required to divest its Spin-Brush business as part of its Gillette acquisition.

13. For the announcement of P&G's purchase of the additional 10 percent of the joint venture, see http://www.cloroxcompany.com/company/news/pr121604-1.html. Other elements of this joint venture were described in interviews with P&G staff on October 20, 2005 and in a visit by Jeff Weedman to my class at Berkeley on May 5, 2005.

14. Lafley's quotation is taken from Suzanne Harrison and Patrick Sullivan, *Einstein in the Boardroom*, 123.

15. Comments from Gus Orphanides, John Tao, Jeff Orens, and anonymous sources are from interviews by the author conducted at Air Products headquarters in Allentown, PA, on October 11, 2005.

16. Collins's quotation and the information on nanotechnology at Air Products is taken from John Teresko's profile of Air Products, "From Confusion to Action," *Industry Week*, September 1, 2005. In the article, Teresko states that Air Products' change process was inspired by the publication of my book *Open Innovation*. While flattering, this attribution is undeserved. As this chapter shows, Air Products' change process started well before the publication of the book.

17. It is ironic but true: companies blessed with significant internal R&D capabilities, which routinely conduct tremendously complex experiments that can run into many millions of dollars, have little or no capability to conduct even simple experiments on the business model supporting that internal R&D.

A great introduction to these issues is Stefan Thomke's book *Experimentation Matters* (Boston: Harvard Business School Press, 2003). If more companies became more adept at experimenting with their business models on a routine basis, there would be less need for a crisis to trigger the experiments that companies like IBM or P&G conducted.

18. While both Ford and GM have been creative in developing sales incentives (e.g., employee pricing, zero percent financing, etc.) and long-term research projects like hydrogen vehicles, neither company seems to be any stronger after many years of cost cutting relative to its competitors. The companies' market shares have declined tremendously, and Toyota is poised to become the largest automotive company in the world in 2007 or 2008. There was a reprieve during the 1990s, thanks to the innovations of the sport-utility vehicle and the minivan, which temporarily boosted American manufacturers' margins and sales. But these innovations were soon copied, and the underlying weaknesses of the U.S. auto industry were again exposed. As of this writing, it is likely that the financial condition of these mainstays of American industrial strength will weaken much further before any lasting improvement is made.

19. Other people at Procter & Gamble who deserve credit for this insight include Nabil Sakkab, who preceded Gil Cloyd as P&G's CTO, and Durk Jager, who preceded A. G. Lafley as CEO.

20. Orphanides' comment comes from interview by author, Air Products Corporate Headquarters, Allentown, PA, on October 11, 2005.

21. These comments come from interview by author with the two men at IBM's Yorktown Heights, NY, research laboratory, October, 7, 2005.

Index

245

About the Author

Henry Chesbrough teaches in the Management of Technology program at UC Berkeley, and is a well-known authority on innovation. His first book, *Open Innovation* (Harvard Business School Press, 2003), articulates a new paradigm for organizing and managing R&D. *Open Innovation* was named one of the best business books of 2003 on National Public Radio's *All Things Considered*, and *Scientific American* named him one of its Top 50 Business and Technology leaders.

Chesbrough is the founding Executive Director of the Center for Open Innovation at the Haas School of Business at the University of California at Berkeley. Previously, he was an assistant professor of business administration and the Class of 1961 Fellow at Harvard Business School. He has a PhD in Business Administration from UC Berkeley, an MBA from Stanford University, and a BA from Yale University, summa cum laude.

Chesbrough's open approach to business comes from his own ten years' experience in the computer industry. He joined a startup disk-drive manufacturer, Quantum Corporation, which outcompeted IBM by selling into OEM channels that IBM eschewed. He helped to found an end-user subsidiary company of Quantum that pioneered the marketing of mass storage products directly to end users. He has advised many leading companies about the benefits of greater openness, including IBM, Procter & Gamble, 3M, Unilever, Philips, Genentech, General Mills, Kimberly Clark, Intel, Hewlett-Packard, EMC, Dell, Microsoft, SAP, and Xerox. He sits on the advisory boards of a number of innovative companies. Recently he founded the Berkeley Innovation Forum, a community of managers that meets to share common problems and effective practices in managing innovation.

His academic work has been published in *Harvard Business Review, California Management Review, Sloan Management Review, Research Policy, Industrial and Corporate Change, Research-Technology Management, Business History Review,* and the *Journal of Evolutionary Economics.* He recently published *Open Innovation: Researching a New Paradigm* (Oxford University Press, 2006) with Wim Vanhaverbeke and Joel West. He is the author of more than twenty case studies on companies in the IT and life sciences sectors, available through Harvard Business School Publishing.

He and his family reside in the San Francisco Bay Area. More information on his work can be found at http://openinnovation.haas .berkeley.edu.